# IT'S HOW
# YOU PLAY
# THE GAME

**ALSO BY BRIAN KILMEADE**

*The Games Do Count*

# IT'S HOW YOU PLAY THE GAME

*The Powerful Sports Moments*
*That Taught Lasting Values*
*to America's Finest*

## BRIAN KILMEADE

## HC

*An Imprint of* HarperCollins*Publishers*

HarperCollins books may be purchased for educational, business, or sales promotional use. For information please write: Special Markets Department, HarperCollins Publishers, 10 East 53rd Street, New York, NY 10022.

FIRST U.S. EDITION

Designed by Richard Ljoenes

Library of Congress Cataloging-in-Publication Data has been applied for.

ISBN: 978-0-06-123726-3
ISBN-10: 0-06-123726-4

07  08  09  10  11  DIX/RRD  10  9  8  7  6  5  4  3  2  1

To my wife, Dawn, son, Bryan,
daughters, Kirstyn and Kaitlyn:

*I love you more than you know.*
*You are truly the A team of nuclear families.*

# CONTENTS

# IT'S HOW
## YOU PLAY
### THE GAME

# INTRODUCTION

My first book, *The Games Do Count,* examined the early sports lives of seventy-three of America's best and brightest, not one of whom was a professional athlete. The letters, e-mails, and requests for autographs poured into my small office at Fox & Friends at an incredible rate. The book made a home for itself on the bestseller list for weeks and sold well for over a year. Why?

Well, judging from the letters, readers bought it because it was the first book that told stories we could all relate to. Most of us don't play professional sports, but when we did play in our childhood, we played with passion and conviction, and our experiences on the field marked us for life. *The Games Do Count* showed that we weren't alone, that many men and women ascribed many of the attributes that made them successful to their early experiences playing sports.

At book signings, I got another clue as to what made the book work. I was asked not just to autograph the book but to make it out to "my coach," "my teammate," "my mom," "my dad," and so on. Obviously, the book helped many players, coaches, and parents connect at all levels, and for that I am honored.

I believe that *It's How You Play the Game* will do what the first book did and more. What do I mean? Well, for starters, what do 5'3" Muggsy Bogues and Pope John Paul II have in common? What can Senator Bob Dole and Mary Lou Retton share the next time they see each other? What can NASCAR'S Jimmie Johnson and NFL Hall of Famer Joe Montana tell you about quitting? The answers to these questions and many more will be answered in the upcoming pages.

*It's How You Play the Game* was not written to teach you how to fight like Lennox Lewis, run like Gale Sayers, or focus like Gary Player. It was written to give you an insider's view of how these great athletes played the game and how playing the game the way they did impacted their lives. Perhaps the lessons they learned can impact yours, too.

A word about how this book is constructed. All the italics are my comments and questions. With the exception of the historic people in the book—Richard Nixon, John Wayne, and George Patton, for instance—everyone was interviewed exclusively for this project.

Amazingly, I found as much in common between Cal Ripken Jr. and his dad as I did between *The View's* Elisabeth Hasselbeck and her dad. How could that be the case? Because the results of the game and the colors of the jersey do not matter. Instead, it's the effort, ethics, and values that emerge from the sport that counts.

Did you know World Cup star Marcelo Balboa got cut from his soccer team by his dad? Did you know racing legend Tony Stewart was all but finished in his sport before he was twenty because of his dad? What they did when they hit those walls is what this book is about.

This is the only book that I know of that attempts to mesh together historical figures like Presidents Teddy Roosevelt and Abe Lincoln and grassroots legends like Coach Ken Carter and Notre Dame's Rudy, as well as all-time greats like Arnold Palmer, George Foreman, and Mia Hamm. This melding of historical figures with contemporary sports heroes works because the era, the sport, and the place do not matter. What does matter is the way they played the game and how they acted and reacted under pressure, when no one was looking.

Can you really learn values, ethics, and morals by taking part in sports? You bet! After ninety separate interviews with an incredibly interesting collection of people, I have come to the conclusion that sports is the best classroom for life. I'm not saying that everyone in this book passed the test every game, every practice—not by a long shot. In fact, many learned the most after ugly moments, like when Ben Crenshaw snapped on the golf course and embarrassed his mom. Stedman Graham, once a demanding, short-tempered basketball teammate and now an ultrasuccessful inspirational businessman, would be the first to admit that if he had never been that fiery hoops star, he would not have become the well-rounded businessman and person he is today.

The inspirational and instructional stories in *It's How You Play the Game* can give parents a valuable resource to fall back on when your kid isn't playing enough, or playing well, or playing hard, or wants to quit the game. This book can also teach you how to relate to your child if he or she is making every all-star team on the planet, or if he or she couldn't make one on your block.

A player reading this book may find inspiration when he or she needs a kick in the pants. It can also be a source of hope when you think you're the only one struggling on the field. You'll see that the trials and tribulations are just part of the process, and that a coach's insightful, caring, and often stern words can flat-out change a player's life. It did for Sean Elliot, George Foreman, Jack Kemp, and countless more.

Why just take in tired axioms, catchphrases, and mottoes when you can expose yourself directly to sports experiences of the men and women you know from politics, business, sports, and history?

What readers won't find in this book are tales of winning. In fact, I was stunned at how little talk there was of championships. Winning is certainly part of their stories, but it's not what makes people who they are. After all, it's not whether you win or lose, it's how you play the game.

—BRIAN KILMEADE

"Keep coming back, and though the world may romp across your spine,
Let every game's end find you still upon the battling line;
For when the One Great Scorer comes to mark against your name,
He writes—not that you won or lost—but how you played the Game."

From the poem *"Alumnus Football"* by Grantland Rice, 1941

# TERRY BRADSHAW

★ NFL HALL OF FAME, 1989

★ 4-TIME SUPER BOWL CHAMPION, 1975, 1976, 1979, 1980, 2-TIME
  SUPER BOWL MVP, 1979, 1980

★ NFL MVP, 1978

★ NFL QUARTERBACK, PITTSBURGH STEELERS, 1970–1983

This isn't nuclear physics, it's a game. How smart do you really
have to be?

—TERRY BRADSHAW

I was a child who needed to be outdoors, and I loved playing any game. My attraction to football was the fascination with throwing this little rubber football. I can't explain it other than to ask, Why do people sing? Why do people dance? Why do people show horses? When I was introduced to a football, it just consumed me. I was determined to make that thing spiral. I didn't know how to do it, but I kept trying. On top of that, every Sunday I watched football with my dad, and I just had to throw the ball like the guys on TV did.

## GET A PLAN

How did I finally learn to throw? Well, in one word, practice. I was living in Iowa and my dad had this huge blanket. I would lay it on this snowbank and throw the ball into the blanket, and the snowbank would absorb the shock and the ball would roll back down.

*And then, in the words of Jim Lampley after George Foreman KO'd Michael Moorer, "It happened!"*

One day, I threw it and it spiraled. To make sure I really had it figured out and that it wasn't just a fluke, I did it a few more times until I was convinced. I remember running into the house and hollering to my mother and asking her to come outside and watch this. She knew I was serious, so she came out and sure enough, I did it again. She knew I thought it was a special moment, and that was good enough for her. I haven't forgotten it, but this is the first time I ever told that story. Here I was, nine years old, and it was the first thing I did well.

## MOST ENJOYABLE TIME

If I had only played college ball and never played a down in pro football, I would have been okay with that. Those years were the most fun because we were free. I made grades, played football, had fun on campus, played in a new stadium. It was just great. You might think I liked college because I did well, but that wasn't the case. It was more about me just being a part of something. It was always about the team. If we lost and I played well, there was nothing good about that. The main reason I liked college was because I loved the coaches and they loved me back. In the pros my coach, Chuck Noll, was a tough love kind of guy, and I couldn't handle it early with the Steelers.

### SUCCESSFUL IN SPORTS EARLY, SUCCESSFUL IN LIFE LATE

I was always five years behind: five years behind in maturity level, five years behind in relationships, five years behind in college. I was clueless to anything that didn't involve me getting to the NFL. I never had good enough grades and I never had a Plan B. I just kept working hard to get what I wanted, with no fallback plan.

### NOW FOR PLAN B

When I was done with pro football, I went right into broadcasting, doing color commentary with Vern Lundquist. The problem was, I didn't know what I was doing, and so I lost all my confidence.

*Wait a minute! Mr. Four Super Bowl rings lost his confidence? Something doesn't compute here.*

I ran into the same problem in the booth as I did in school, and that was remembering names. I couldn't match up faces and players, and I was all but overwhelmed. I know it when I study it, but I kind of lose the words when the light comes on. Even today, you don't see me getting into many specifics with players, because I don't know their numbers.

### PRESSURE? BRING IT ON!

Nevertheless, I enjoyed broadcasting, and I learned to do well under pressure. I take a great event and then downsize it on my mind, so I can relax. I tried hypnotism. I even used buzzwords like *relax, confidence,* and *concentrate.* Eventually, I learned to release all that energy in a positive way.

*Well, he wasn't known as the best clutch quarterback in history for nothing.*

At one point, I was so relaxed I almost fell asleep in the locker room before games. But all this helps me perform in front of an audience. The first thing I do is to strip people of their titles, and then I strip the event of its importance. I convince myself that they're my friends and they're not better than me. And then, when I've stripped them of all that, they're just like me, so I'm out there talking to a bunch of me's. It settles me down and makes it easier to talk to them.

## PERSONALLY SPEAKING

I lost all my money twice, I've been divorced three times, I've been called stupid and dumb. And that's just a starting point! I learn what I need to know to get comfortable at it, and the whole time I'm going full speed ahead. I hear the critics, but it doesn't stop me. It's never stopped me and it never will.

## MY WRAP

*Just when you thought you knew him as a guy who had it all, you learn that no one has it all. Terry, especially, never had nor will he have it easy. The important thing is, he kept moving forward, kept learning, and kept working, and the end result is two distinct Hall of Fame careers, one as a player and now one as a broadcaster. Who knows, his next frontier might just be acting. I saw* Failure to Launch *and ol' Terry was* great. *Of course he'd be the last to acknowledge it, but he'd always be appreciative that you said it.*

# EVANDER HOLYFIELD

★ **4-TIME HEAVYWEIGHT BOXING CHAMPION**

★ **BRONZE MEDAL, 1984 OLYMPICS**

★ **NATIONAL GOLDEN GLOVES CHAMPION, 1980**

Hurting people is my business.

—SUGAR RAY ROBINSON,
*welterweight and middleweight champion*

I always dreamed big. During the 1976 Olympics I was watching a feature on the Spinks brothers, Leon and Michael. The whole world loved Sugar Ray Leonard and Howard Davis, but I related to those two brothers. They were from the projects, as was I. I looked at those two and thought, "*It is possible.*" Later, I found out it would take faith and hard work, but at the time, seeing them made it all seem possible and within reach.

## PUT ME IN, COACH

I had a hard time getting my football coach to put me in the game because I was a little guy and he didn't know the size of my heart. My first goal was to play for the Atlanta Falcons, but the problem was that when I got to high school I just could not get on the field. I was 110 pounds, and although the coaches thought I was good enough to make the team, they didn't think I was good enough to actually get in the game. I prepared each week like I was going to start, and each week I watched from the bench. It was incredibly painful and frustrating. I knew I would get a chance at some point and that I had to be ready when it came.

Can you imagine having a team that could not use Evander Holyfield?

Tired of waiting, I just stood up at practice one day and said, "Coach, put me in at middle linebacker."

He said, "Evander, those guys are one hundred and ninety pounds and you're too small."

I repeated, "Coach, put me in, and if the runner gets by me I'll go back to the bench. But if I make the tackle, you keep me in. Deal?"

He nodded, and I went in. The opposing coach called the play and it was a screen pass to the fullback, who was lumbering right toward me, and I stopped him cold at the line. The sideline cheered. My coach said, "Good hit, now take a seat." I watched from the bench for the rest of the season. I did not miss a practice. I was never late to a drill, and still I did not get another chance until the fourth quarter of the last game of the season. I played linebacker and made about eight tackles in twelve plays. The coach came over to me at the end of the game and said, "I didn't know you could play like that. See you next year." Too late. I proved I could play, but I was done with football. It was time to become a boxer.

## MOM, LET ME QUIT!

I wanted to quit boxing because there was this one fighter, Caesar Colin, who beat me twice for the Junior Olympic title. I just could not beat this guy, so I decided I wanted to quit. I told my mom and she said, "No, you will not quit because you're not doing well or because you're frustrated. If I let you quit because you didn't beat this kid, you'll be quitting things your whole life." She told me most people quit things when they're not doing well. "Beat this kid, win the division, then come to me and we'll do whatever you want with boxing." So, at twelve, I got another shot at Caesar and I beat him. I went running home, told my mom, and she said, "Okay, Evander, now you can quit boxing."

"Are you crazy?" I said. "Why would I quit after beating my toughest opponent?"

Only then did I realize what she was trying to teach me. It was official—I had an irreplaceable life lesson.

## FINAL THOUGHTS

I am so thankful that my mom did not intercede and tell my brothers and sisters to go easy on me because I was the youngest of nine. I am also grateful that I had that frustration early on in football, because it was the first time I hit some turmoil, and I did not quit.

Life is not fair. My brothers didn't let me win, my football coach didn't put me in, and that's just what life is all about. I looked at the boxing ring as a testing ground to show my will and how I handled pressure. I welcomed the chance to test myself every day. I won every fight for eight years and then I lost. Everyone was looking to see how I handled defeat. I didn't blame anyone else and I didn't quit. Instead, I studied how I lost. And almost every time I came back and won, and I was grateful, because after every loss I was forced to become a better fighter.

Life is about making adjustments. When I talk to my kids or go to a school to talk to kids, I let them know that I might be a champion, but sometimes I lost. Even as an amateur, I lost eleven times. But I won one hundred sixty-five times. After each loss, I went back, studied my mistakes, and came back a better fighter, as you should come back in whatever you are dealing with in life.

## MY WRAP

*Evander Holyfield's name may be synonymous with courage, tenacity, and class, but his path to the title was anything but easy. Like too many of us, he was benched unjustly because a coach didn't give him a chance. Remember, if that coach had picked up on Evander's heart, he probably would have stayed with football. He might only have been just another linebacker instead of one of the finest fighters in boxing history. He certainly would not have been as rich as he is today, nor would he have become globally famous. On a personal note, his victory over Mike Tyson the first time was one of the most inspiring sporting events I have ever seen. Tyson had all his opponents cowering. Holyfield was coming off a horrible performance, and he outboxed the most feared man on the planet. It's hard to match that drama. To me, Evander would make a great broadcast color commentator or a phenomenal trainer.*

# JERRY WEST

★ **LAKERS GENERAL MANAGER: 4 NBA TITLES**

★ **NBA HALL OF FAME, 1980**

★ **AVERAGED 29.1 POINTS PER GAME IN 153 NBA PLAYOFF GAMES**

★ **LOS ANGELES LAKERS, 1960–1974, WINNING CHAMPIONSHIP IN 1972**

★ **CAPTAIN, U.S. OLYMPIC GOLD MEDAL–WINNING BASKETBALL TEAM, 1960**

★ **NCAA CHAMPION, TOURNAMENT MVP, 1959**

Basketball is like war in that offensive weapons are developed first, and it always takes a while for the defense to catch up.

—RED AUERBACH,
*legendary Boston Celtics coach and general manager*

I started playing basketball when I was seven or eight on a dirt court in front of my house, and I've never really stopped. As a kid I was bored and needed something that would challenge me. It was great, because I could see progress, even at a young age.

## THE POWER OF THE MIND

As I got older, I would always go to watch the upperclassmen play at what would eventually be my high school. My mind would wander and I would see myself playing, visualizing myself making the last shot to win a game. It was the beginning of goal-setting for me, and it let me know the power of my mind and of my imagination. I have always been a bit of a loner and this was a game where you could work alone, drilling yourself on the things that would make you a better player. I didn't have a great home life, so I could just go outside and escape.

## JUNIOR HIGH

I wanted to play in junior high, but I was so small and timid, I just couldn't seem to get myself on the court. But that didn't stop me from playing and trying to become a better player. I wasn't even dreaming about going pro, I just wanted to get better. And I did begin to improve.

## HIGH SCHOOL HEIGHT

I got to high school and suddenly I just shot up, going from this little kid to a tall, gangly kid. I was brought up from JV to varsity and people began to notice. I went down one last time to play JV and then rejoin the varsity, but I broke my foot. Through it all, I wouldn't stop playing. In fact, I broke about six casts because I kept shooting baskets.

## REVEALING MR. CLUTCH

I started my junior year, but our team was not very good. I was tall at 6'3", but I only weighed one hundred sixty pounds. My senior year was the coming-out party. We won the state championships. I set scoring and rebounding records in the tournament, and my life changed. I was recruited from all over the country, but I only wanted to go to West Virginia University, so that's where I went.

What mattered to me most was going out to that dirt court, imagining hitting the last shot, playing as if I were on every team, playing every position on the court. I wanted the ball at the end of every game. I was never nervous, because countless times in my head I already imagined what it was like to take and hit the last shot. Whether it was in my team gym, on my dirt court, or through the coat hanger hanging on my door, I always felt like I had been in pressure situations because of my vivid imagination.

### ALMOST WINNING IT ALL

No one took losing as hard as Jerry West.

—CHICK HEARN,
*longtime Lakers announcer*

At times, I almost feel like my career was a failure because we did not win more NBA titles than we did. There were at least two times where I know we were the better team and we didn't win. Ironically, when I played my worst, we won, and that was against the New York Knicks. I noticed at that time how differently people treated you after you win, and it soured me because I saw how fickle people are. I know what it's like to fail, but I also know what it's like to get up and try to achieve your goal. As much as I take pride in trying, I am still not over those losses, even today.

### FINAL THOUGHTS

I've never been driven by my ego. I have been driven by my desire.

—JERRY WEST

My imagination brought me this success, and I also had an internal belief that I was going to be the best player I could be. Fear of failure drove me. I just couldn't accept losing, nor did I want to visualize it. My family never pushed me to do anything in sports. It all came from me. I didn't do it to be rich or famous or to go pro. I just wanted to be the best player I could be. Even today, I do not like seeing my name in the paper and do not like talking about myself. I do not take myself seriously; I take what I do seriously.

My goals are simple, and the last three years they have been exactly the same: try to be a better person than I was last year and give more of myself personally. I want the people I work with to know I care about them, and I want to see them move forward. That's what I try to do.

## FINAL, FINAL THOUGHTS

Sports teaches you more about life than any other job you could have. Sports, like life, is a marathon, not a sprint. You know you're going to have bad injuries and terrible losses, but you have to keep marching and keep your eye on the big picture. As an athlete, I've learned to be resilient and to overcome adversity. And if you ever lose your competitive edge, just hang it up. I never worked for money, only to compete and to win.

## MY WRAP

*What Jerry West has done in basketball is just about more then anyone else in NBA history. West was one of the best players, coaches, and general managers of all time. He's another example for those who think you can make it on talent alone. It's all about the practice time alone and the intensity you show to make yourself better. If Jerry West says he learns about life and sports every day, don't you ever think school's out and that you know enough.*

# BYRON "WHIZZER" WHITE

- ★ ASSOCIATE SUPREME COURT JUSTICE, 1962–1993
- ★ DECORATED WWII NAVY VETERAN
- ★ NFL RUSHING LEADER, 1938–1940, DETROIT LIONS
- ★ GRADUATED #1, YALE LAW SCHOOL; RHODES SCHOLARSHIP TO OXFORD
- ★ VALEDICTORIAN/ALL-AMERICAN RUNNING BACK, UNIVERSITY OF COLORADO

A team that has character doesn't need stimulation.

—TOM LANDRY,
*legendary Dallas Cowboys coach*

AS TOLD TO ME BY AUTHOR DENNIS J. HUTCHINSON,
AUTHOR OF *THE MAN WHO ONCE WAS WHIZZER WHITE*

Byron "Whizzer" White spent a lifetime divorcing his professional life from his other lives, both private and athletic. There were, however, a couple of times when he referred to athletic experiences that had shaped the attitudes and values he held throughout his life.

One example was when he was sent as deputy attorney general to Alabama to deal with the conflict between the Ku Klux Klan and a group of blacks during the Freedom Rides crisis in 1961. The attorney general was very concerned that there was going to be a race riot on his watch. As the conflict mounted to a climax, White was sitting on a bench at the Air Force base and he said to his colleague, John Doar, "This is how you get tested." Doar later told me that by the look on his face at that moment he could tell that he was flashing back to his days in football when the odds were against him and yet he would forge forward anyway. It seemed to me he was thinking back to the odds on a riot between the Klan and this group of black churchgoers. He knew the police were not going to stop it or provide adequate protection, so he had to put together his own defense system.

### "COURT OF LAW AND LAW OF THE JUNGLE"

The second moment that stood out for White happened the first year he turned pro for the NFL's Pittsburgh Pirates (later the Steelers.) He went into the league as the highest-paid player ever, earning a whopping $15,800 a year. This fact, as you might imagine, made him a target for every player in the league, barring his own teammates, of course. And every time he ran the ball, he'd get an extra beating.

*White explained how he was told to handle the resentment to then assistant attorney general William Orrick. Orrick was having problems with some backstabbing associates and sought out White for advice:*

That happened to me when I started playing professional football. I was with the Pirates and after the whistle was blown, they were kicking me and I asked the coach, what'll I do? "Wait till you catch one of the out-of-bounds passes after the whistle's blown," he said, "and then you kick him there (insert imagination) and kick him in the face and make sure everyone sees you. It'll cost the team twenty-five yards, but I'll be able to keep you for a couple of seasons."

That's what Byron did—it did cost the team twenty-five yards—but he never had any trouble after that.

Essentially, according to his brother, Byron had the same attitude when he studied as when he played. He seemed to always demonstrate total self-control and total focus. His brother went on to tell me that when Byron studied, no dogs barked.

Byron loved to watch and follow the games. He would flip open the *USA Today* sports page before anything else. He told me he loved to see first class athletes in their prime being tested. No doubt that was because he tested himself athletically and he wanted to see others tested the same way.

## HIS OWN VIEW OF HIS SPORTS LIFE

Byron viewed playing sports as his duty because he felt it was a job he was entrusted with and not something that would glorify him. I can't know if he worked harder on the game than on the books, but I do know he worked extremely hard at everything he did.

His teammate, Art Unger, on Byron White the football player:

> Byron would have been just as happy, or maybe even he would have preferred it, if he played with twenty-one other players in an empty stadium—no fans, no coaches, no refs. Football was a personal challenge, something he could use to test his own limits. He just hated the stuff that happened before or after the whistle. He also became a hero to generations of student athletes all across the country. He actually left pro football to go Oxford for a year.

*Can you imagine someone doing that today?*

## BROTHERLY IMPACT

The person who made the biggest impact on Byron was his older brother. He was a Rhodes Scholar before Byron and he went on to become an extremely distinguished physician.

Although he carried so many similar qualities into both his legal and his sports careers, Byron hated to mix the two. Once, when a teammate paid a visit to the Justice Department, he greeted Byron with his customary, "Hi Whizzer."

Byron's sheepish reply was, "Geez, don't call me that here."

He didn't want to mix business with pleasure.

## MY WRAP

*Byron White may have led the most complete, successful, diverse life that I've ever heard of. Through his rise in pro football to his days with the Kennedys to the Supreme Court, he was not only the best, but he was humble. He played because he loved to compete, and not for the fame it might bring him. He was the personification of the student athlete, as well as a true rarity: a dominant figure in two distinct careers. It's amazing how his days in football taught him how to handle his critics and step up under duress.*

# DANIELLE GREEN

★ U.S. ARMY SPECIALIST, 571ST MILITARY POLICE COMPANY

★ NOTRE DAME BASKETBALL PLAYER, 1995–2000

The only discipline that lasts is self-discipline.

—BUM PHILLIPS,
*former NFL head coach*

My mom smoked crack for the majority of my childhood, and my dad was never around. I was an only child and I looked at sports as an outlet, a way to get me out of the inner city. I just had to break the cycle in my family. I was home alone a lot and on Saturdays I used to watch Notre Dame football every week. Even then, I thought there must be something special about that place. I focused on my grades and athletics so I wouldn't end up like my mom or anyone else in my family. I would spend all my free time going to parks with my semi-flat basketball. I decided I was going to go to Notre Dame.

The first organized ball I played was in the sixth grade. At the time I was just a rebounder. I was big, but I didn't know how to shoot. Fortunately, I had a coach who knew the game and drilled us on the fundamentals. The more I learned, the more I wanted to learn. I remember getting a bike and telling my mom I was just going riding. What I really did was go down to the high school. It was a dangerous place—there were drug shootings there—but I didn't care. It wasn't like I felt comfortable sitting in my house. I would get in the game, playing against twelve- and thirteen-year-olds, and handle myself pretty well. Soon I was standing out, even among kids much older than me, and I started feeling good about myself for the first time.

### FIRST MOMENT

One day, I was working on my game at the high school when a guy who'd been staring at me walked up to me and said, "Wow, if you ever work on your right hand you could be a great ballplayer." It might have been the first praise I got in anything at any time. I'm not even sure if he knew anything about the game, but I knew he was right. As a natural lefty, I had never taken the time to learn how to use my other hand.

### HIGH SCHOOL

I went to Roosevelt High School in Chicago, and I have to say I put the team on the map. They had not won much until I got there, but once I did the team started to soar. My freshman year we went undefeated, and then we started playing big-time schools. It was great for me because I played three different positions and was pretty much the entire offense. The highlight had to be the conference finals. I scored forty-two points with seventeen rebounds as a freshman and we won. I was never the type of player who got in anyone's face. I just wanted to stay strong in school and get to Notre Dame. I knew my teammates were not putting the extra work in, but I still had to keep myself on track. I could feel that my life had taken a turn. I could have done whatever I wanted, because I had nobody looking out for me, but I was always motivated to make something of myself.

## FROM STAR TO SUB

I got to Notre Dame and for some reason I stopped fighting. It was almost like I forgot how to work hard. I was also not the best player on the team, like I'd been in Chicago. I was coming off the bench, which was hard, so I averaged only one point a game and I wanted to leave. Over the summer, I got ready to roll and then I tore my Achilles tendon. I was devastated. My junior year, I started to show some promise and Coach McGraw started to ride me hard. I thought she was picking on me. I didn't get it. Now I know she knew I could be great and didn't see the drive and dedication from me that were necessary to actually *be* great. We fought constantly throughout my five years at the college.

Looking back, I was just not prepared for the size and speed of the opponents playing against me every game. And I never took the time to develop my right hand. I started off great my last two years and ended up on and off the bench. Part of it was that I just could not handle not being the star. I guess you could say I was selfish. I became possessed by stats. I didn't understand the team concept. I thought I could be the star and here I was, the fourth or fifth option. In fact, I almost didn't get my fifth year granted because my coach thought I was a cancer. I guess, in a way, I was, despite the fact that I still averaged twelve points a game. In retrospect, I was looking for a mother figure to give me some type of support and my coach was not used to playing that role.

## A LIFE-CHANGING MOVE

I graduated college and tried out for the Detroit Shock of the WNBA. I was the last cut and was let go. I tried teaching, which I liked, but I wanted a challenge, so I decided to join the army. I signed up on September 16, 2002. I knew the Iraq war was coming, but I still did it. I loved the uniform and thought serving my country would be the noblest thing. By 2004, I was in Iraq. I'd been in Iraq four months when a rocket-propelled grenade hit me in my thigh, and I lost my left arm below the elbow. I had to relearn everything, because my left hand was dominant. Brushing my teeth, combing my hair, and trying to tie my shoes were all a challenge.

## "NOW I FINALLY *HAVE* TO WORK MY RIGHT HAND."

What helped me get through this wound and the loss of my arm was Coach Muffin McGraw. After being hit, I was shipped out to Germany, and it just so happened that a friend of Coach McGraw's was also in the hospital. She contacted my coach, who called me in Germany. I told her, "Coach, you always told me to use my right hand and now I have to. In reality, it was her toughness that got me ready for this challenge. She was tougher than any drill sergeant I'd ever had. She demanded more of me and let me know I could do anything. Also, I think playing on the streets of Chicago hardened me, allowing me to put up with the pain. It helped me with the adjustment to my new situation.

## READY FOR THIS?

As I came to and realized what had happened, I thought about life without my hand. On my left hand, which had been blown off, were my wedding rings. They weren't even paid for yet, and now they were gone. Well, word got out about my rings and the soldiers in my unit, my new team, went back out, and found my hand and the rings, risking their lives to do so. I thought, *"How lucky am I to have people in my corner like that?"* I instantly eliminated feeling sorry for myself.

## FINAL THOUGHTS

Without basketball I would have been in a gang. I used to think they were so cool and they all had so much money. And maybe I'd have had a few babies. I just hate to think about life without the game. Now I'm out of the army, but I don't regret a thing. It changed me for the better. I'm in grad school now, and yes, my right hand is getting better every day. Now I just have to worry about overusing it.

**THE WRAP**

*If Danielle's life isn't worthy of a movie, I don't know whose life is. Let's all think twice before we ever complain about our circumstances again. Here's a situation where sports were used to save a life and carve out a future that had been destined for disaster. Again, no sports career is perfect, and it took a grown-up Danielle to recognize that the collegiate Danielle was selfish. What's most important is that she learned from her experience. I think she'll make one heck of a college coach someday.*

# BILL YOAST

★ HIGH SCHOOL FOOTBALL COACH FEATURED IN THE MOVIE
*REMEMBER THE TITANS* (PORTRAYED BY WILL PATTON)

Most coaches study the films when they lose. I study them
when we win—to see if I can figure out what I did right.

—BEAR BRYANT,
*legendary University of Alabama football coach*

I grew up in Florence, Alabama. When I played football it was really about having fun. Coaching was more personal back then, not the way it is today. I remember my coach saying, "If you're not having fun out there, then I don't want to see you back here tomorrow." Don't get me wrong, we were worked hard, but winning wasn't everything. If we lost, we didn't hang our heads. Our coach wouldn't stand for that. He was the dominant figure in my sports life and his philosophy is how I approached the game for the rest of my life.

## WHEN PLAYER BECOMES TEACHER

I like to think I've learned a lot from all my coaches and mentors along the way. The person who I learned the most from wasn't a peer or a pro but a player, Gary Bertier (played by Ryan Hurst in the movie). As you know, if you watched *Remember the Titans,* Gary was one of the best players I ever had. But, tragically, he was in an accident and lost the use of his legs. Instead of getting down, he started participating in the wheelchair Olympics. He never felt bad for himself, and he battled for the rights of the handicapped.

## WHAT I LEARNED FROM COACH BOONE

What I learned from Coach Boone (portrayed by Denzel Washington) is that no matter your approach to sports or life, you can still work together and learn from each other. He was a yeller, an in-your-face type coach, and I wasn't. We wanted the same thing, we just took a different route to the goal. Over time, he learned that my way also got the job done. I wanted to learn as much as possible about his offense, because it was unique and dominant. But I also shared my philosophy on defense with him, and I'm sure he'd tell you I taught him a thing or two.

## MY WAY

I learned that you do not treat all players the same because each player is different. Remember Peete? He was getting screamed at on offense by

Coach Boone and came over to defense. I made him a starter and he made all-state, but only because I knew you can't scream at a guy like him. Coach Boone also noticed I was too soft on a few guys and so he gave them a kick in the rear end. We had an amazing relationship.

## LIFE WITHOUT SPORTS?

If I didn't play sports I would never have gotten out of high school. I was a poor kid with little confidence. I was looking for an identity and I just dove headfirst into sports—all sports—including basketball and track. I learned that it's not how hard you fall but how high you bounce back up. Winning is important, but failure is a great teacher. Play the best you can, but if you don't win, don't have any regrets. Coach Herman Boone would fight me on this, but I just think life is so much more important than a game.

## ON RACE

The more I coach, the more I realize that all kids, race and creed aside, have the same dreams, the same hopes. I've never coached a kid I didn't love and I think that'll never change.

## ON THE MOVIE

The season portrayed was 1971 and here we are all these years later in Hollywood watching our lives being rehearsed and perfected before our eyes. I thought about how far both Herman and I have come. Me, a white kid from the cotton fields of Alabama. Herm, a black kid from the tobacco fields of North Carolina. Now we're in the Rose Bowl, hanging out with Denzel Washington. It was just an incredible experience.

## HOW DO YOU PLAY THE GAME?

Prepare yourself well. Play with intensity. Be ready to change and adjust. Brace yourself for adversity, because it's part of life. You are not born with character; it's something you work at and learn. Here's how I put it best to a group of all-stars I was asked to address: "Most of you are here because you are talented, but if you got here on your talent alone, this could be the last game you play. What will take you to the top is attitude, loyalty and integrity, and those are things you can't measure."

Most coaches go easy on their blue-chippers because they don't want to upset them. But those are the ones you have to push the hardest because they have to learn the work ethic that will eventually take them to the top in whatever they choose to do. If you get blue-chippers with the plugger attitude there's no telling what your team can do. I know because I just had a plugger who came to my house to tell me he just became a member of the Navy's Flight Demonstration Squadron, the Blue Angels. And I just crossed paths with another plugger who had been the assistant secretary of state under Bush Forty-one.

**MY WRAP**

*Coach Yoast also told me that his evolution as a coach went from loving his blue-chippers to worshipping his pluggers. Often it's the pluggers who enjoy the most success in life because they don't know what it's like to have anything come easy. They find their thing and attack it like only a plugger can, and then they come out on top by a mile. I loved the movie and this story, don't you?*

# JOEY CHEEK

★ *TIME* MAGAZINE'S 100 MOST INFLUENTIAL PEOPLE

★ OLYMPIC GOLD MEDAL, SPEED SKATING, 500 METERS; SILVER MEDAL, 1,000 METERS, 2006

★ WORLD SPRINT CHAMPION, 2006

★ OLYMPIC BRONZE MEDAL, SPEED SKATING, 1,000 METERS, 2002

Sports, more than any other activity, has proven that a truly democratic society owes the individual nothing more than an opportunity.

—KEITH JACKSON,
*former sportscaster*

I knew, from the first moment I could stand, that I wanted to race. I started in-line skating at ten and yes, I always wanted to win. Even as a kid, I was ultracompetitive. I was a fairly unstable, high-strung kid who was thirsty to be a champion at something. Skating looked to be my ticket because I was so good and so quick. I made it to the national level of the competition my first year on skates and I thought I had really arrived. The next year, I moved up an age group and got shellacked, never even making it out of the regional competitions. And now I know why. Not only was I small for my age, but I didn't train. I thought I could just go out there and win, but now I knew those days were over.

## THE KEY MOMENT

The first time I was beaten was a sobering experience for me, and it was the last time I would let myself get too high or low after any season or race. It's a principle I still keep with me fifteen years later. I never let myself get too big an ego because the universe will always find a way to knock you back down. It was also the last time I would race without having outworked almost everyone in training. Today, at the international level, I am not close to being the best athlete on the ice, but few work harder than me and I'm convinced that's the reason I've found success. I also try to outthink my opponents. I want to be the smartest athlete in every race.

## SMALL GOALS, BIG RESULTS

I have three technical goals that I work on every practice session. Number one: keep my back round; number two: head level; number 3: making my stroke as my right leg comes back to my body. And because I have trained so hard and in such detail, every big race feels like just another practice.

*Okay, we're not speed skaters, but it's good to know that this elite skater is still working on fundamentals. So, next time you think you're too good to hit balls off a tee, or too old to kick a ball off a wall, think of Joey Cheek.*

## IT HASN'T ALWAYS BEEN THIS WAY . . .

The crucial moment for me came leading up to and just after the 2002 Olympics. Up until 2002, the problem was that my self-worth was tied into how I skated. I was so set on qualifying for and winning the Olympics that I would live and breathe it night and day. I'd go to bed after a great session and worry that I couldn't repeat it tomorrow. If I had a bad practice, I thought it was the end of the world. I was not sleeping well at night. And if you don't sleep and you worry all the time, your performance suffers. Somehow I managed to make the team and won a bronze medal, but I did not get any enjoyment out of it. I made a pledge to myself that either I quit or I untie my sense of self-worth from how I skate. And so winning the bronze was worth it because of the very valuable lesson I learned.

## CARE LESS, WIN MORE

During the next four years, I promised myself that I'd forget about the outcome and focus on the process, the training, to see where it took me. And let me tell you, I loved the last four years. I worked through injuries and it just didn't bother me. I refused to worry about things I could not control. I just set my own individual goals.

*Which weren't shabby, considering that Joey has three Olympic medals.*

I found out how much I really loved the sport.

It wasn't easy, though. I had my mom and a sports psychologist helping me with this transformation. But it paid off. The year after the 2002 Olympics, I won medals at every World Cup event and the world championships, including my first world gold medal. What was best for me, though, was that I was smiling the whole time. It was fun to skate and, oh yeah, it was great to win.

## THE OLYMPICS

I had a great Olympic experience in 2006 and I was thrilled with the result: one gold and one silver. But I have to say I really would have enjoyed it even without the hardware. I know I would not be the person I am today if I had not been an athlete. I proved to myself I could accomplish great things. But what you do with those accomplishments really defines you. If you want to live on, you have to try to make a lasting impact, and for me that's all about helping people to better their lives.

**MY WRAP**

*But let's face it, the fact that he won has given him the chance to spread the wealth: the real-world gold. He knows his notoriety, fame, and recognition will fade, but while it was fresh in the minds of the world, he wanted to use it to help someone else. And so he rallied his supporters to match what he donated to help Africans—he's particularly committed to the Save Darfur Coalition, which is concerned about genocide in the Darfur region of Sudan. What he donated is the prize money the United States Olympic Committee gives to every medal winner. For taking the gold he got twenty-five thousand dollars for his bronze he received fifteen thousand dollars; and he quickly handed it over to an athlete-driven humanitarian organization called Right to Play. His donation prompted others to match it to the tune of over four hundred thousand dollars. Joey's not exactly living in the past. Right after the games he was accepted to Princeton and began classes in the fall of 2006. He finally has balance and, in my mind, as bright a future as any young man in America.*

# DWAYNE "THE ROCK" JOHNSON

- ★ STAR OF *THE SCORPION KING, THE RUNDOWN, WALKING TALL, BE COOL, DOOM,* AND *GRIDIRON GANG*
- ★ RECORD 7-TIME WWF/E CHAMPION
- ★ WWE HALL OF FAME INDUCTEE
- ★ DEFENSIVE END, UNIVERSITY OF MIAMI

Sport is cut and dried. You always know when you succeed . . . You are not an actor; you don't wonder "did my performance go down all right?" You've lost.

—STEVE DAVIS,
*British snooker player*

I was at home one day when I was a teenager and saw myself on TV featured on "Crime Stoppers." I hunted down a cop I knew and asked for help. He said, "I'll help you on one condition. I want you to stop screwing up and go out for football." I agreed, and he set me up with a team. This was really when I began to focus on the game in a committed way. Before that, when I was eight, I'd played some Pop Warner football, but I almost never got in the game because I was just awful. But that was okay with me, because I was scared to get in there, scared I was going to get hurt.

*The Rock, scared? Now that's hard to believe.*

When I went back to the game at fourteen, I was bigger and faster and I loved the contact. That's when I really started to love the game. Until then, I was making bad decisions, fighting, hanging out with the wrong guys. I thought I knew it all, but in reality, I knew nothing.

### "ON THE MOVE"

When we moved to Pennsylvania I came across a great coach—Jody Cwick. He brought me to another level, and soon I was getting college offers from across the country, and it was all because he saw potential in me as a person and as a player. I played tight end and the defensive line, and I enjoyed every moment of it.

### ON TO COLLEGE: THE UNIVERSITY OF MIAMI

I lucked out when I ran into another coach, Bob Karmelowicz, who took an extraordinary interest in me. When he recruited me he said that he had a term for my style of play that I still think about today. He said, "Dwayne, you have great upper-body violence. You control the player and back him up and it's something I can't teach." As promised, he spent time with me at school working on harnessing that talent.

### COMPETING TO BE A PRO

I got off to a strong start battling future number-one pick Russell Maryland for a starting spot on the defensive line, and I ended up being second string as a freshman. Until, that is, I dislocated my shoulder, and then I was done for the year. Just like that.

**HIS WORLD TURNS UPSIDE-DOWN**

I couldn't handle not playing, so I left school, just walked out. And then I got a call from head coach Dennis Erickson during Christmas break. He said, "Get your butt back to school." He found out that I had a .5 GPA. Ouch! He ripped into me, telling me that I had one last chance to get my grades up or I was out of school for good—that I'd lose my scholarship, that I'd be through at Miami. He gave me a note card and I had to get every professor to sign it, saying that I was really in class.

*You don't have to use conjecture to see how sports helped Dwayne. No school without it!*

This changed my life. I went from being almost kicked out of school to academic captain of the team. It was my job to help other guys to stay eligible. I learned how to study in school, how to set up my schedule, and how to budget my time, and these are things I still use today in my life as an actor and as a parent.

**ON A ROLL**

By the time I was a junior, I was a preseason All-American, and that's when the Hurricanes brought in a guy named Warren Sapp, who was to become a number-one draft pick of the Tampa Bay Buccaneers. Well, there went my chance of starting. But I was okay with that because I knew he was flat-out better than I was. I would see Sapp and linebacker Ray Lewis and I'd think, "*These guys are great,*" and I had to find a way to get great myself.

**THE QUEST**

And so I would challenge myself every day. But during one practice I heard a "pop" while doing a pass rush drill. I became numb. I couldn't move. It turned out that I had ruptured three discs. I should have sat out the season, but I got shot up and played the whole year and let me tell you, I played badly. My senior year came and went, and when the draft came I thought for sure that I'd be drafted. But I wasn't. I was crushed, because at Miami almost everyone seemed to get drafted by the NFL. Not me. I was a pick of Calgary, of the Canadian Football League.

## DREAM DERAILED

What's worse then being forced to play in Canada? That's easy. Being cut. And guess what, I deserved to be cut. I was not able to play at that level. I had all these dreams and aspirations from the age of sixteen—I was going to play for the Giants—and I just did not achieve my objective. To not have it happen, well, it's humbling, but it set me up for success in the next phase of my life and career, which leads me to the mantra I adhere to today.

### "BECAUSE I HAVE BEEN HUNGRY, I WILL NEVER BE FULL"

I had to move home with my parents. I was married, but we couldn't afford to live together because my parents lived in a two-room apartment. My former teammates were buying cars and houses for their parents and I was living with mine. It hurt a lot and it didn't help that I had just seven bucks in my pocket.

## DONE WITH FOOTBALL

Today, I'm proud that I never felt sorry for myself. Instead, I just closed the door on the game and came up with another plan. My plan? To become a pro wrestler. I managed to get a contract and before I knew it, I was on my way. In the beginning, I wrestled in front of twenty to two hundred people a night while living on the road, driving fifteen hundred miles a week.

I was making forty bucks a show in the smallest of small towns. I wrestled in venues ranging from used car dealerships to huge barns. I was cut by fans, hit with batteries, but none of that was important because I was learning the business. Even when I was called up by Vince McMahon and signed by the WWE, I was being booed out of the building, even when I had the Intercontinental Belt. And I hated it.

## SO, HOW DID THIS CHAPTER CLOSE?

With an injury. Yes, I tore the PCL (posterior crucite ligament) in my knee and they wrote me off. They took away my belt and I found myself feeling like a failure again, living at home, rehabbing alone, not knowing what I would do.

## THE LAST GASP

The big moment for me came with a phone call from wrestling executive Jim Ross. He said, "Dwayne, I have an idea for you. Come back to the ring as a bad guy. A heel." I said okay, and the rest is history. I became one of the most famous and successful wrestlers, which then allowed me to get into the movies.

## FULL CIRCLE

Now I'm in Hollywood, playing the fictional role that I hoped to live in real life, first as a coach of wayward kids in the box office smash hit *Gridiron Gang* and now as the star quarterback going to the Super Bowl and getting the MVP in a movie called *The Game Plan*. I get to live out my dreams on the big screen, and I just love it.

## FINAL THOUGHT

I feel blessed that I fell short, or dare I say failed, at my first dream, because it made me hungry and even more determined while at the same time letting me know that there are no guarantees to success. I appreciate my success so much more now because I did not have it early and because I worked so hard to get where I am at today. I pursue both the game and life relentlessly and with passion, and let the chips fall where they may.

## MY WRAP

*I know what you are thinking: How did a kid who was getting into so much trouble get so many people in his corner at crucial times in his life? The answer: respect. In the Polynesian culture they teach respect for adults, and even when he was being arrested and failing school, coaches, teachers, and cops saw how respectful he acted. Even during his darkest hour, they were able to see the good and the potential in Dwayne. Think about that when you see one person seemingly getting all the breaks. Perhaps he or she has earned it through good old-fashioned manners and deference to others. Dwayne showed promise and expressed appreciation in the midst of turmoil. One final observation: I have* met and observed Dwayne in a number of different settings and as you know, he's enormously popular, and yet I've never seen him brush someone off or not express appreciation after a photo or autograph request. He does not demonstrate the megastar ego because, in my opinion, he knows what it's like to be on the bench, to be cut from the team. His life exemplifies the classic career audible. He had a flexible game plan, wrestling. What's yours?

# DOROTHY HAMILL

★ ICE CAPADES HEADLINER, 1977–84

★ OLYMPIC GOLD MEDAL, FIGURE SKATING, 1976

★ WORLD CHAMPION, 1976

★ 3-TIME UNITED STATES NATIONAL CHAMPION, 1974–76

You can take an ol' mule and run him and feed him and train him and get him in the best shape of his life, but you ain't going to win the Kentucky Derby.

—PEPPER MARTIN,
*baseball player, St. Louis Cardinals*

I was a bad student and I was painfully shy growing up. One day my mom and I decided to try something new, figure skating. So, at eight years old, I started skating and entered my first competition at nine. It was held at the Wollmann Rink in New York City's Central Park, and I felt like such an outsider. All the other girls had these great outfits and knew each other and I was alone, wearing a dress that went down past my knees. Somehow I managed to come in second, and I had so much fun. I can't explain what a thrill it was and at that moment a new world opened up for me. I found one place in my life where I felt comfortable.

### THE OLYMPICS: NOW, THAT WOULD BE COOL!

I kept skating without a goal until 1967 when, as a Christmas present, my parents gave me tickets to watch Peggy Fleming compete at the Olympic qualifiers. I thought, "*Wow, she is brilliant,*" and it was the first time I realized how cool the Olympics would be, without thinking I was Olympic material, of course.

### TRAINING, YES. SACRIFICE, NO

As I got older, I got more serious, training whenever I could, even going up to Lake Placid in the summer when my friends would go off to summer camp. Going to Lake Placid was not a sacrifice for me because I just loved it. In fact, I did not sacrifice my childhood to skating. I was a shy kid who felt comfortable on the ice. My parents never pushed me. If anything, I pushed them to keep me out there on the ice. I would practice fours a day and fought for more, not less.

### BREAKTHROUGH

A great moment for me was winning the national championship in the novice division at twelve. At fourteen came the real test—I began to compete with the big guns. And when I placed fifth in Oklahoma, it was the first time I realized I could compete with the best in the country. People started looking at me as an Olympic hopeful, but I never bought into that.

## OLYMPICS

I began to realize that making the 1976 U.S. Olympic team was a possibility, though at nineteen actually winning never crossed my mind. But I did make the team, and to my amazement, earned the Gold . . . it seemed like the whole world watched! I skated, laughed, and cried, and my life has never been the same. To this day I don't feel worthy of it.

## IN ADDITION

Skating was where I first learned about winning and losing in life. It's where I learned the ethic of hard work, which I thrived on then and today. I learned to love being pushed by my coaches and I learned to respond in kind. I'm not saying I would not have learned all these things about life and myself without skating, but I don't know anywhere else I could have picked it all up as quickly and as cleanly. For some reason, anywhere else in my life when people would challenge me or tell me I was not good enough, I would take it personally and more than likely agree that I did not measure up.

Overall, there is nothing better than knowing you have done your homework, put in the hours studying, and done your best. It far outweighs the feeling of winning or losing.

## THE WRAP

*Here's an example of a young girl needing sports to help her belong rather than needing them to grow. Dorothy loved to practice, possibly because staying on the ice beat leaving it, where she felt totally out of her element. Winning the Olympics forced her out of her shell and made her grow up almost before the world's eyes. Now, finally, Dorothy Hamill can feel like she has life on and off the ice.*

# EDGAR MARTINEZ

★ **NAMED THIRD BASEMAN ON MLB'S LATINO LEGENDS TEAM, 2005**

★ **THE MLB AWARD GIVEN ANNUALLY FOR THE BEST DESIGNATED HITTER IS NAMED THE EDGAR MARTINEZ AWARD**

★ **7-TIME AL ALL-STAR**

★ **MAJOR LEAGUE BASEBALL THIRD BASEMAN AND DESIGNATED HITTER, SEATTLE MARINERS, 1987–2004**

Most of what I know about style I learned from Roberto Clemente.

—JOHN SAYLES,
*writer*

My first memory of sports was watching Roberto Clemente of the Pittsburgh Pirates play in the World Series. I just couldn't believe a player from Puerto Rico like me was so great. I just always remember wanting to be like him, so I found a league and a team and I started playing like mad and never stopped.

*In 2004, Martinez was given the Roberto Clemente Award for being the player who best combined outstanding baseball skills with devoted work in the community.*

It turned out that I was one of those kids who had a lot of ability. I could catch and hit right away, and that just made me want to play more. My teens were tough years because I not only hard to work hard to improve my baseball skills, but I also had to have a job in order to continue to go to school. I'd go to school at night, from 6:00 PM to 10:00 PM, and then I went to work at a factory that made all kinds of things from 11:00 PM to 7:00 AM. And three times a week, I played baseball. So I guess you could say that I worked at the factory so that I would be able to play ball. That's how important it was to me. But as important as baseball was to me, I did not phone it in at my job. I took tremendous pride in my effort. By the end, I was the factory supervisor, with twelve people under me. If it weren't for baseball, I probably would have worked my way up the ladder at that place.

## THE COACH WHO MATTERED

I had one coach who truly made a difference for me. His name was Jose Antonio and he was a local guy who knew the game and cared about the kids. He would pick us up, take us to the ball field, and work us out. I guess he saw something in me, because he did all this for free, just to see if I had what it took to get to the majors.

## JUST BE PATIENT, MOM AND DAD

At fifteen, my parents sat me down and said, "Edgar, please go get a job. Give up this idea of playing baseball and stop pursuing sports." It was tough, because I knew I wanted to be a pro, but I didn't know if I had what it took. I ended up making everyone happy by going to college for a year, but I still continued to play ball to keep the dream alive.

## THE KEY FOR ANYONE, IN OR OUT OF SPORTS

I was always obedient, disciplined, respectful, and hard-working. Whether I made it to the majors or not, I learned early on that these four qualities would be present in any job I did. Many of us have ability, but unless you have the principles and values to go along with it, that doesn't necessarily translate into success.

## A WINNER NEVER QUITS, BUT SOMETIMES THE TEMPTATION IS THERE

*At every level, he was the best hitter on his team, but once he was in the Mariners farm system, being the best wasn't good enough. Can you believe it took an injury for this minor league star to get a chance at the majors?*

In 1987, I won the Triple-A batting title (*he hit .360*), and it looked like I was going to get my chance to play in the big leagues. But it didn't happen. In 1988, they saw me play in the spring and still kept me down. I thought it was over. I didn't know to ask for a trade. I just thought, "Maybe I'll never get my chance in the majors." But I knew I couldn't do much better than .360, and yet the team had shown no interest in keeping me on the roster. I just could not figure out what was happening, because I had never been ignored before. One player, Carmelo Martinez, told me to stick it out. Hard to believe, but I was up and down in 1989, too, but in 1990 I got my chance. The Seattle third baseman, Jim Presley, got hurt and they had to turn to me. We started to win. I hit .300 and never looked back.

## SOMETIMES YOU HAVE TO SAY ENOUGH IS ENOUGH

Look, I know I wasn't perfect. I've charged the mound, and I've gotten tossed out of games. And you know what, I don't regret it, because each time I did it either I was being thrown at or one of my teammates was being dusted. It was up to me to let the other team know that I wouldn't stand for it. And guess what, the result was that teams completely changed the way they were pitching to me.

*And why does he remain a fan favorite?*

I think it's because the fans knew that I came to play, that I cared about the team first, and that I really love the city of Seattle.

## WHY ME? WHY NOT?

I'm just an average guy who loved his job and went at it with passion. I was going to work hard and do well at whatever I chose because I was not looking for a free ride. I was ready to work. Even today, I'm just as motivated to make my company, Caribbean Apparel, as successful as the Seattle Mariners.

## MY WRAP

*Edgar's being modest about his work ethic and demeanor. He was incredibly consistent, and it's not an exaggeration to say that his play, especially in the 1995 postseason, saved baseball in Seattle. When the stars of the team— Randy Johnson, A-Rod, and Ken Griffey Jr.—went elsewhere, he stuck around. That's what builds a bond with the fans. That's how the game should be played.*

# BOB DOLE

★ **REPUBLICAN NOMINEE FOR PRESIDENT, 1996**

★ **SENATE MAJORITY LEADER 1985–87, 1995–96**

★ **U. S. SENATOR FROM KANSAS, 1969–1996**

You can learn little from victory. You can learn everything from defeat.

—CHRISTY MATHEWSON,
*Hall of Fame pitcher, New York Giants*

I was a standout player in many sports in high school—football, basketball, and running the quarter mile—but when I got to the University of Kansas, it was tough to realize that I wasn't good enough to play football or basketball, which left only track. Still, I had that work ethic to carry me through.

It was my high school football coach, George Baxter, who got the most out of me. He saw a 6'1", one hundred eighty-pound solid end who was a good guy, a whip who could be a leader for him on the field. I also had a basketball coach, Harold Elliot, who saw some leadership qualities in me and let me know I would go a long way in whatever I did. What made this special for me was that he knew I wasn't going to be an All-American basketball player, but he still took an interest in me as a person. It meant a lot to me then and it means a lot to me now.

### OFF TO WAR

I was planning to go off to World War II and come back and carry on in sports in some capacity and then go to medical school. After being wounded in Italy, I knew that dream was done.

*Bob was shot in the upper right back and bled for nine hours, and wound up with an arm that was almost unrecognizable.*

Going home and living with my arm in that condition was never easy, but throughout my rehab I was able to call on my sports work ethic to get me through. At Walter Reed Medical Center, they would tell me to do eight repetitions of a particular exercise and I would do ten. They'd say do fifteen and I'd do twenty. This is the same way I was taught to compete in sports: always overdeliver. My training in track and football helped make the brutal transition more tolerable, and it made the rehab fly by.

### FINAL THOUGHT

I would like to have been a better player, but in the end, simply playing sports made me a better man. Let me also add that my days as an athlete helped make me a tireless campaigner. I loved going long days on little sleep. One time, in 1996, at the age of seventy-three, I did ninety-six straight hours of campaigning. My staff was amazed, but I wasn't. I was a track guy who constantly trained, and if there is one thing I knew I'd always have, it was endurance.

**MY WRAP**

*Senator Dole was a much better athlete than he confessed to, and it's the drive that he freely admits that helped get him up out of that hospital bed. He is a shining example of why you play and how you should compete to get maximum results.*

## ANGELO DUNDEE

★ HALL OF FAME BOXING TRAINER OF FIFTEEN WORLD CHAMPIONS, INCLUDING MUHAMMAD ALI AND SUGAR RAY LEONARD

If you ever get belted and see three fighters through a haze, go after the one in the middle. That's what ruined me—I went after the two guys on the end.

—MAX BAER,

*1934 heavyweight champion*

I was an aircraft inspector before World War II, but all I knew was prop planes. When I got back from the military, the planes all became jets, which is when I called my brother, Chris, and asked if I could work in his boxing gym. It was 1948, and Chris had sixteen fighters and he let me come down and be the bucket boy for the greatest trainers in the world: Charlie Goldman, Jimmy White, and Ray Arcel. They opened up their years of experience to me. I worked out of Stillman's Gym, where my brother had his office, and I even had a studio couch, where I basically lived.

Those men taught me so much, but first and foremost, they taught me to help people and share with others what you know so they can grow. I learned about people from them, how to deal with different people in different ways. There were different buttons you had to push with everyone. I also learned that you can't teach desire.

### MUHAMMAD ALI

I had to make Ali feel like he innovated everything, even down to the slightest punch. Later in his career, he used to say, "Angelo don't train me." That's exactly what I wanted him to think. I just wanted him to do the things he needed to do to get better. What people should know about Ali is that he's a wonderful human being, a better person than he was a fighter. He would come to Christmas parties at my house and go play with the kids.

### SUGAR RAY LEONARD

What I loved about Angelo was that he had a way of motivating without ever showing panic. He was always in control. He'd express urgency, but I never saw him desperate. He was always a calming figure in my corner.

—SUGAR RAY LEONARD

His first fight, Ray shows up in Baltimore with his name on his robe. I said, "Ray, you don't need a name on your robe for these people. If they don't know who you are, they don't know boxing." It was all part of getting my kids to believe in themselves.

## WILLIE PASTRANO

> Notice who is in the locker room after you lose, not after you win.
> —ANGELO DUNDEE

One night, Willie walked out into the ring and just got his butt kicked. He went back into the locker room and it was almost empty. Before the fight, you couldn't even take a deep breath without hitting someone. I said, "Willie, take a look around. These are your friends. This is exactly like life. Who's around you when the chips are down? Those are people you can count on."

*Angelo, that didn't bother you, not getting credit?*

I never wanted credit. You have to know your role in life, and for me it's staying on the fringe. In fact, the first few years with Muhammad, they thought I was a mute. I never talked. In public, it's all about the fighter. I made it a point to always be available to the media because the media makes and breaks fighters. Before Ali, these fighters never talked. I wanted them out there all the time. I don't say "I." It's always "we." I just tried to make each fighter the best they could be and that's all a coach or a trainer should do. I don't create. I just try to better what a guy already has.

## WORKING WITH THE STARS

I worked with Will Smith for his role in *Ali*, and Russell Crowe for *Cinderella Man*. These guys were like fighters. They worked like dogs. They're driven. They were great students who just oozed talent and drive. Russell ran in the bush for his roadwork and road his bike to the gym. He had a gym on his farm. This guy was just like a fighter. And he became a good fighter. I watched that movie and I saw Braddock. I treated Smith just like he wanted to be treated, and he wanted to be treated not like a star but like a fighter. And he could have been a fighter. Crowe could have been a fighter, too. Just think about the careers I could have ruined if I'd gotten to these guys at the right time!

**MY WRAP**

*Like many coaches, Angelo is selfless. Some people like to say it's not about them, but with Angelo it's really true. He didn't need the spotlight and he doesn't need it now. He loves competing, but treats everybody the same way, and that's respectfully. He's the first one to say that he can't put toughness and drive into a person if it's not there to begin with. Too bad he never hooked up with Mike Tyson.*

# COACH KEN CARTER

★ MOVIE BASED ON HIS LIFE, *COACH CARTER*; PORTRAYED BY SAMUEL L. JACKSON

★ HEAD COACH, RICHMOND HIGH SCHOOL (RICHMOND, CA) BASKETBALL TEAM, 1997–2002

★ GRADUATED AS RICHMOND HIGH'S ALL-TIME LEADING BASKETBALL SCORER

★ RICHMOND HIGH SCHOOL—LETTERED IN FOUR SPORTS PLAYER

Coaching is nothing more than eliminating mistakes before you get fired.

—LOU HOLTZ,
*legendary college football coach*

Everything I am and everything I am not I owe to sports. Practice does not make you perfect. Practice makes you better. Average is not good enough.

—COACH CARTER

I had an advantage over other high school athletes because my brother was a high school All-American. I saw how hard he worked, and I knew success came with a price before I even walked into the gym. I had a real good high school career and left as the team's all-time scoring leader. What I'm most proud of is being the first one at practice and the first one to finish his running, and if I wasn't playing well, my coaches would pull me over and discuss it with me and not treat me like the other kids. They did this because I think they recognized how motivated I was and they saw the drive I had. Why was I like that? In my family with seven sisters, you learned to take orders well, making it very similar to a team situation.

*What Coach Carter didn't say is that Richmond, California, was the most dangerous city in the state. Fifty percent of the kids who entered high school there did not graduate, and students were eighty times more likely to go to jail than to college. He was way ahead of the game, defying the odds just by going to college.*

## THE MOMENT: 1999

> Coach Carter cares about you, it's as simple as that. He won't let us fail.
>
> —WAYNE OLIVER,
> *former player*

I put my sporting goods business on hold to take over the Richmond High basketball program. First, I instituted a nineteen-point pledge that each player had to sign. They could stay in the program if, among other things, they kept their grades at a minimum 2.3 average, attended all classes, and sat in the front row. The next season, our team was undefeated, going 13-0, but as I made rounds to check on my team grades, I discovered that fifteen of the forty-five players in our freshman, JV, and varsity program were failing. These kids were so caught up with their success on the court that they would not do their classroom work and were not responding to my crackdown. So I locked the gym. I saw too many high school legends left behind by life and I did not want to see that happen to these kids. I made them study every day until their grades came up. In the meantime, we forfeited two league games and made national news for putting school first. The team ended up 19–5, losing in the second round of the state tournament. Many people were upset, but the message came through to those kids and they knew they had only themselves to blame for the suspension.

### KEY

I tell kids they have to make up their minds early that they won't do drugs and that they will go to college, in order to be successful. If you make up your mind you will not cut class and will not smoke, when those things come in your path you will walk away. It's the people without defined goals who are the most vulnerable.

### THE MOVIE

We had so much national press that we had studios and producers calling me to do the movie on our team. Once the deal was done, I made sure to be on the set every day because I had to make sure the story was accurately told. I fought to get this done the right way. Even though they were making a movie, I felt as if I were still coaching. Samuel L. Jackson was a true pro. He didn't stumble on one line, and he spent time with my family and my team. By the end, he was actually coaching in a live game. In the movie, they were actually playing and they had to make their shots, with no cutting away. I made them get players and not actors to play my team and the other teams, and that, I think, was key to the movie's success.

### THE MAGIC OF WRITING THINGS DOWN

One day when I was eight years old, I came home from school and found my mom crying because she couldn't pay the bills. I went into the other room and wrote down the following on a white piece of paper: "Mom, one day they will make a movie about me and I will pay off all your bills and you will never have to cry ever again." It took thirty-five years, but it happened, and I still have that note today.

## FINAL THOUGHTS

The most important things in life are scores. Think about your grade point average and your credit scores. Those are numbers by which we are judged. Sports is decided by numbers and it's how our team is often defined. So how is sports different from life? It isn't. It *is* life. I always tell people to enjoy the process of earning a degree or buying a house, because often the process is more enjoyable than attaining the actual goal.

## MY WRAP

*You know from the first word that comes out of his mouth why a movie was made about Coach Carter. Myriad messages can be taken from his story, from writing down goals to the need for a strong family structure. Anyone who can book this coach as a keynote speaker should. He's got an important message that parents and players need to hear.*

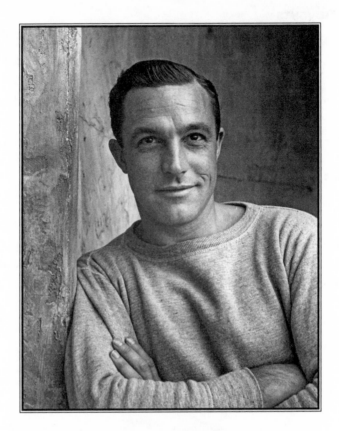

# GENE KELLY

★ **LIFETIME ACHIEVEMENT AWARD FROM THE AMERICAN FILM INSTITUTE, 1985**

★ **AWARDED THE *LÉGION D'HONNEUR* BY THE FRENCH GOVERNMENT, 1960**

★ **APPEARED IN THE FILMS *AN AMERICAN IN PARIS* AND *SINGIN' IN THE RAIN*, AMONG MANY OTHERS**

★ **HOLLYWOOD DIRECTOR AND PRODUCER**

★ **BROADWAY AND HOLLYWOOD DANCER, ACTOR, AND SINGER**

The man who complains about the way the ball bounces is likely the one who dropped it.

—LOU HOLTZ,
*legendary college football coach*

Gene Kelly was not only a great athlete but also a sports nut. He had a theory on dance that the dancer should approach it as an athlete. Growing up a poor Catholic boy in Pittsburgh, he dreamed of playing shortstop for the Pirates. But what he really loved was hockey. Had there been more than a handful of teams in the NHL at the time he grew up, he might have had a shot as a pro.

Kelly's mom was very ambitious and used to clean houses so his sisters could take dance classes. Gene walked them to class, to make sure they got there on time and safely. As the story goes, he was sitting, watching class, and he decided to hop up and dance up a storm. Before that, it was thought to be a sissy thing to do, but after trying it, he was hooked. He was a small but tough Irish kid, bright in school and great at all sports, but the idea of dancing had not occurred to him. Of course, back then there weren't organized leagues to play in. Instead, he played in pickup games in backyards and ball fields. Soon he and his sisters started doing some shows, and people began to notice. There was a lot of talk about this tough, good-looking street kid who danced with a muscularity about him. They saw that he brought something different to dance. He always thought of it as a sport and approached it the same way. No doubt, that swagger and physicality he learned in sports created his superstardom in dance.

On his movie sets, Gene was known to put together all kinds of games, from softball to volleyball. Everyone had to play. He felt it bonded them together and broke up the boredom. One day he went for a ball on the volleyball court and broke his ankle. By that time, Fred Astaire had been in semiretirement, but Gene called him in to fill in for him on the movie he was rehearsing with Judy Garland. He did it, and it sparked something in Fred and he danced for years after that. (The film was *Easter Parade*.)

He thought dance was a real sport and compared the moves on the field to anything on stage. Lynn Swann was his favorite football player because he loved the body control Swann showed. Dad always believed he was a better hockey player than dancer. His dad used to ice over the backyard and he and his brother used to play ice hockey all the time. My dad was also a gymnast. Even when he was sixty-five he could sit in a chair, pull his legs up, and go into a handstand. He retired when he did because he didn't think people wanted to see old boxers or dancers past their prime. He felt like an athlete—you have your time, then you're done. What made retirement easier for him was that he loved choreographing shows and he was quite good at it.

My dad loved testing himself and just competing. It wasn't all about winning with him, it was about doing everything he could to win. I watched him play tennis and I didn't think he even knew the rules. Somehow he'd figure these games out, and he ended being an excellent player who used his head.

## SELLING HIS SPORT

Gene and his brother got into a lot of fights. Later, when he started taking dance classes, they got into more fights. That's why he wanted dance to be accepted as a sport. He wasn't into the top hat kind of thing that Fred Astaire was. He wanted everybody to know he could dance, but he did it without compromising himself. Later, he would become friends with some of the biggest names in sports, including Mickey Mantle.

## SUCCESS WITHOUT SPORTS

It's impossible to separate sports from Gene and his success in dance. He was great at any sport he tried, and I can't imagine him being the person he was without sports. If he didn't find the game, I have a hunch the game would have found him.

## MY WRAP

*Sports helped the great Gene Kelly create a whole new brand of dance. He was a masculine, muscular guy who made dance cool for a whole generation of kids and parents. He also seemed to be a guy who would be honest with him-self about his skills. If he thought he would have been a better hockey player than dancer, I'll trust his judgment. I'm sure he made the right choice, though, because very rarely do dancers lose teeth or get tossed over the boards from behind.*

# KERRI STRUG

★ **OLYMPIC GOLD MEDAL, U.S. WOMEN'S GYMNASTICS TEAM, 1996**

★ **OLYMPIC BRONZE MEDAL, U.S. WOMEN'S GYMNASTICS TEAM, 1992**

There has never been a great athlete who died not knowing what pain is.

—BILL BRADLEY,
*former NBA player and U.S. senator*

My sister was training for gymnastics with Bela Karolyi. When I was five years old I went to visit her. I guess I had the body and the attitude, and this caught his eye. His attention and encouragement led to my taking lessons for two hours a day, five days a week with the University of Arizona coach. The sessions were one-on-one and they were short but intense. I finally hooked up with Bela himself when I was thirteen. Up until that time, it was all fun for me, but when I joined Bela it became much more like a job. The best way to describe it was enjoyable, but intense.

### HER FIRST COMPETITION

It was just a small competition in my own gym, but I was so nervous. I wanted to do well so badly and so I performed badly. It wasn't exactly a disaster, but I didn't even do as well as I did in workouts. This became a pattern for me. I would make small mistakes and Bela would become very frustrated with me. This pattern lasted until the 1996 Olympics. What made it worse is that all my teammates would perform better in meets than in practice and for me it seemed to be the opposite. I knew I was working as hard as my teammates—maybe harder—and yet I just was not winning. I was good, but rarely did I come in first.

It wasn't until my last year, when I thought that I had nothing to lose, that I started focusing on me instead of everyone else. I had to look closely at what I could control and what I couldn't. In the past, I had been worried about how good my teammates, like Dominique Moceanu and Shannon Miller, did. But that all changed, and just in time for the Olympics.

### MIND-SET—MY BEST IS GOOD ENOUGH

Gymnastics was every important to me, but it wasn't everything, and that's because my parents would not allow it to be that way. Unlike most of my teammates, my parents never pushed me to keep going. They were more interested in my going to school. I didn't need gymnastics to win my parents' love. They loved me whether I won or not. The pressure came from me and only me.

It all came together for me when I saw the finish line at the 1996 Olympics. I knew that win or lose, I was done competing as a gymnast after that. In a sense, that relaxed me. Looking back, I think I wanted to win so badly that I was afraid to lose and as a result, I didn't do well. After that, I began to win a lot, mainly because I came to realize that I could only control my own performance, and not other gymnasts.'

## THE VAULT: 1996

I hit my first vault but landed short and jammed my left ankle (afterward, doctors told me I had torn two ligaments). I felt tremendous pain but I had no second thoughts about doing two vaults. I had thirty seconds between vaults. I looked at Bela, lined it up, and did it. At the time, we all thought the team needed that vault to get the gold. In the end, I could have skipped the second vault and still have gotten the gold, but that's not the point. Bela made me always do twice as much as I thought I needed to do at practice. My feeling always was, "Hey, I hit twenty perfect vaults, now let me go home," but he wouldn't let me. In the end, I was so used to this kind of thinking that I was on automatic pilot. Bela was a tough coach and he knew I was in pain, but this was the Olympics. I knew that if I didn't do the second vault I would have questioned myself the rest of my life. And so I did it and as I did I wasn't thinking about the pain, I was only thinking about nailing it.

## WHAT I KNOW

Before Atlanta, I thought that with hard work and belief in yourself, you could do whatever you want and do it well. After the vault, these beliefs became a reality because I proved them to myself. If you work hard, things may not turn out exactly the way you envision them, but they will turn out well. Now, unlike my days in gymnastics, I've come to enjoy the journey to my goals at least as much if not more than reaching the goals themselves. Before learning these lessons, I simply didn't have the balance one needs in life. Then it was all about sports, and so often I didn't enjoy life because I was so fixated on making it to first place. Fortunately, that's not the way it is for me today.

## NOW

I went back to get my master's from Stanford and then I worked at the Treasury Department. Now I'm with Juvenile Justice. I like what I'm doing, but I still haven't found that second passion. I'll do all I can to find it. Whatever happens, you can count on one thing: Kerri Strug will give all she has to become a success.

I love that we won the gold in the Olympics, but I have to be known for more than just being a world-class athlete; I have to be more than one-dimensional. And that's what drives me today.

## MY WRAP

*Kerri's story is a perfect example of how a person can make life harder than it has to be. Strug's moment came when she decided to stop trying to control everyone else's performance and let her own performance speak for itself. So many have said that when you have fun you stop caring about winning and then, guess what, you start winning. Kerri always worked hard, and once she removed the pressure she put on herself she became one of the most celebrated athletes in U.S. Olympic history.*

# JIMMIE JOHNSON

★ NEXTEL CUP CHAMPION, 2006

★ DAYTONA 500 WINNER, 2006

★ RANKS SECOND AMONG ALL ACTIVE DRIVERS WITH AN AVERAGE OF 4.5 WINS A SEASON

★ 1998 ASA ROOKIE OF THE YEAR

Your car moves faster than you can think.

—EDDIE SACHS,

*race car driver*

My first sports moment of impact happened in my mid-teens. I'd been racing motorcycles at the local dirt tracks and I was ranked in the top three in the country in my age bracket. At the age of thirteen, I even had a contract with Suzuki—they sponsored my bike. To keep in training, I was running miles during the week and eating carefully, because my dad would tell me someone else was racing or training that day and I'd panic and go hit the road.

It wasn't long before I got burned out. I didn't take advantage of my dirt bike opportunities. I lost the Suzuki contract, had a few injuries, and took a year off. My parents never pressured me and just let me go back to being a kid. I spent that year looking back, beating myself up for not doing more with all the opportunities I had.

*Is he kidding? Disappointed at the age of fourteen for not taking advantage of the rare opportunities coming his way? This makes the pressures of being a Hollywood child star sound routine!*

## BEING A KID: OVERRATED

I was so hard on myself that the pressure that sidelined me began to reverse and motivate me. I decided to get back in, but to leave dirt bikes behind and go to off road racing. Unless you were worth millions, you had to woo sponsors, and I was getting good with that part of the business. I went into overdrive. I was able to go up through the ranks of off-road to asphalt racing and establish myself.

## OVERCOMING THE FEAR FACTOR

Rick Johnson was a world champion in motorcross and he was like a mentor to me. It was amazing, having a guy whom I looked up to as a teammate. He helped me with fear, teaching me when I could take risks and go to the other side of that line and when I shouldn't. He taught me to get all I could out of a car and out of the equipment without taking too much of a risk. It got to the point where when I'm racing, I don't ever think I'm going to crash until the moment before I hit the wall.

## HOW DOES ALL THIS HELP?

I have lived a long time in my thirty years. Financially, it's likely that I'm set for life, even if it all ends today. But if it does, I've learned so much about life by just trying to be great in racing. I've found out that you have to have total commitment to achieve your goals. I packed my stuff and moved from San Diego to North Carolina with nowhere to stay because I had to see if I could make it in racing. I didn't want to have any regrets. I sacrificed friendships and family for months just to pursue my goal, and that was making it in NASCAR. I didn't have the drive and commitment as a teenager, and it all came apart. I was not going to make the same mistake again. It was a risk, but without laying it all out there I just don't see how I could possibly have been successful in anything.

## NOT QUITE THERE YET

I have won races, trophies, and money, but what was left was winning it all. I did that in 2006 and it was better than I ever imagined.

## MY WRAP

*It's amazing to think that a thirty-year-old already thinks he's gotten a second chance at a great racing career. And it's interesting to note that Jimmie quit racing at fourteen, the exact age when 70 percent of Americans quit sports. But he went back to what he loved, to what he was meant to do, and he says it was his personal drive that pushed him back in. Keep in mind that his parents never pushed him; the decision was his to bail and his to come back. Perhaps this should give us a clue as to how to deal with kids who have similar decisions to make.*

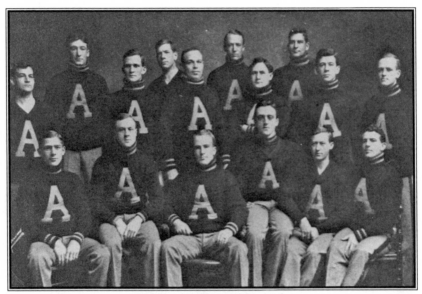

*Patton, top row, second from right*

# GENERAL GEORGE S. PATTON

★ ACHIEVED RANK OF 4-STAR GENERAL, 1945

★ AWARDED 12 U.S. MEDALS, INCLUDING A DISTINGUISHED SERVICE
  CROSS WITH ONE OAK LEAF CLUSTER, SILVER STAR WITH ONE OAK
  LEAF CLUSTER, BRONZE STAR, AND PURPLE HEART

★ AWARDED 11 FOREIGN MEDALS

★ 1912 STOCKHOLM SUMMER OLYMPIAN, MODERN PENTATHLON

★ U.S. ARMY, 1909–1945

Accept the challenges so that you can feel the exhilaration of
victory.

—GEORGE S. PATTON

When my mom told the general (her father) that she wanted to marry my father, the general was not happy. But he relented, mainly because he knew my father was a great athlete. My dad was a terrific horseback rider and swimmer, and my grandfather loved that. I would play three sports a year my whole life, just because my mom and dad came from families where sports were deemed a vital part of your development. By the way, in military circles the name Totten gets as much respect as Patton. The Tottens have a rich history of military accomplishments and acts of heroism.

My grandfather looked at sports as a way to prepare for battle. As a kid, he decided he wanted to be a general. He learned that soldiering was a physical job and you had to be in great shape. He believed competition brought out the best in people, and he strived to do his best. If his best was good enough, he'd win. He pursued athletics because he thought it made him a better soldier by helping him deal with fear and by staying fit. He saw football as a true test of strength and courage. When he tried it himself he found that although he loved it, he wasn't very good at it. His way of coping was to try harder. That's where the broken bones came in.

General Patton was quite slim in his college days. He was a second-string, one hundred fifty-pound linebacker, but he couldn't seem to stay on the field. Throughout his four years of football at West Point, he kept getting hurt. This was pre-pad football. All they wore for protection were these big, bulky sweaters and leather helmets. The way I understand it, he tried so hard to make the first team that he kept breaking bones and would end up in a cast, rarely making an appearance on the field.

FROM STANLEY HERSHON'S BIOGRAPHY, *GENERAL PATTON: A SOLDIER'S LIFE* (HARPERCOLLINS PUBLISHERS)

In the fall of 1906, Patton busied himself with football. A substitute end on the West Point team, he had his biggest moment not during a game but during a practice against the varsity before the Yale contest. When Patton's block against the starting right tackle enabled the scrubs to gain fifteen yards, he drew compliments. But generally, he played so poorly that he was not even in the practice games.

His senior year, Patton was intent on breaking into the lineup and starting for West Point. Unfortunately, his fourth year looked like it was going the way of the other three. In this letter to his future wife, Bea, you can feel his passion to play:

> We are in one H__ of a fix for football men. We have no tackle and not a single scrub . . . I am about as heavy as any body on the scrubs and I can only just get 160 in undress it is awful. . . . When the Navy got us last year I swore a great swear that come what would I would do anything in my power to beat them this year and if the destined way is to be a center, a center I will be. If only we can win. It may sound silly but my thoughts are anything but light on the subject.

At the end of September he sent Bea terrible news.

> We had a very hard practice Monday and I did myself proud until the last down when I broke my small bone of my left arm. I am rather put over (so is the bone) as I was going in the game Saturday. Also, it has kept me awake and continues to do so. I will be out five weeks and that gives me four more weeks of this, my last season. I have already devised a brace so hope to get in again.

## ACCORDING TO JAMIE TOTTEN

My grandfather's real talent was in track. He held a couple of records at West Point in the hurdles. At the West Point field day in 1908, he won the 120-yard and 220-yard hurdles, and also placed second in the 220-yard dash.

In the 1912 Olympics in Stockholm, Sweden, he competed in the first-ever pentathlon. In those days, only those in the military were allowed to compete. His approach to track and to sports in general can be best summarized in a letter he sent and one he received from his dad. "Old Blood and Guts" wasn't all about winning, not by a long shot.

*In June 1905, when Young George tripped over a hurdle and finished last instead of a sure second in a race, his father wrote:*

> Your letter of the 3rd came to-day and I can't tell you how my soul sympathized with you in your defeat in the hurdle race—but it was only because I knew how much you had set your heart on success. It is a good thing to be ambitious and to strive mightily to win in every contest in which you engage; but you must school yourself to meet defeat and failure without bitterness and to take your comfort in having striven worthily and done your best.

On May 10, 1912, Patton got word that he was to represent the army in the modern pentathlon at the Olympic games in Stockholm. The events consisted of pistol shooting, swimming 300 meters, fencing, cross-country steeplechase (three miles on horseback), and a cross-country run of two and a half miles. He finished fifth overall, and if it weren't for an early case of nerves during the pistol shooting, he would have won a medal. The most noteworthy moment was perhaps his last run. He left the stadium in first place and came back in third and almost fainted at the finish. One newspaper marveled at his fine finish. Also of note was that he bested the French army fencing champ and did poorest in an event (shooting) he should have been best at. He still managed to right his ship and come back.

*This letter from his dad points out how special his son's performance was. He wrote from California to Sweden, where George hung around after the games:*

> It is too bad you are not able to be here just now. It would give you a graphic idea of that elusive thing called "fame." I can hardly walk a block in town without meeting an enthusiast who rings my hand and says "that son of yours is surely a wonder" or some similar expression.

*As for the poor showing in the shooting event, his father wrote:*

In shooting it is universally conceded you must have been doped—or your cartridges tampered with.

Great is fame—few enjoy it while alive. You ought to come home before it is forgotten.

**BY THE WAY**

Being General Patton's grandson is a lofty legacy to live up to. I am just as proud of my father's family as I am of my mother's. George's grandfather was a major force in Los Angeles, and he was the one who named Big Bear Lake because he killed so many grizzlies there. They were really the founding family of California and literally owned Pasadena. That was a lot to live up to. I managed to cut my own wake in the military, rising to the rank of colonel.

**MY WRAP**

*The legendary general experienced it all in sports. The one constant was his passion and drive. He went as hard at the sport in which he did not excel (football) as in the ones he did (track, shooting, and fencing.) The compassionate letters from his dad should give heart to those who think a kind, caring dad raises a weak kid. That wasn't the case for me, nor was it for George Patton. When I read about Patton's life in football, I was inspired to write this book and determined to make historic figures a part of it.*

# SIMON COWELL

★ JUDGE OF TELEVISION'S *POP IDOL, AMERICAN IDOL,* AND *THE X FACTOR*

★ AUTHOR OF *I DON'T MEAN TO BE RUDE, BUT . . .*

★ EXECUTIVE PRODUCER, *SO YOU THINK YOU'VE GOT TALENT*

If it's a cliché to say athletics build character as well as muscle, then I subscribe to the cliché.

—GERALD R. FORD,
*former U.S. President*

What I liked about sports was the freedom I got as a kid—I was not monitored when I played sports. We were able to get away with a lot. Unlike today here in America, when we played parents did not attend races or matches, and local reporters were not there. We were just able to compete, work out our differences, and leave. Recently, over in the UK, some idiot from the Labour government said we should not let our kids compete in sports. It's absurd. If you stop kids from competing in sports, they will not compete in life, and life is a competition.

For me, playing sports meant soccer and running track. I was quite a fast runner. I ran the 100, 200, 400 and 800. I wasn't a bad soccer player. I played inside right, because I wanted to be [1960s and 1970s UK soccer legend] George Best.

What I didn't like about sports taught me a lot about what I wanted to do the rest of my life, and that's to be the best. I played hard, worked hard, and was not, shall we say, near the best. That bothered me. Okay, it didn't just bother me, I hated it. I hated playing hard and not winning, not dominating, and in some cases not even being good. It was at that point that I made the decision, the declaration really, that if I can't be the best, I just wouldn't do it. I concluded that it's better to be the best at one or two things than average at loads of things.

*I guess you could call this Simon's anti-Nike moment.*

After much initial success racing on raw natural ability, I got my comeuppance. I lined up to run 400 meters for the first time, thinking, "Well, I've never run it before, but I'll figure it out." After all, I knew how to win the 800, so what was the difference? Well, I learned quick when the gun went off and this guy takes off out of the blocks in a full sprint and there I am in my *Chariots of Fire* pace. All I could think of was, "What the hell is he doing?" I picked up the pace and finished second—a *bad* second, by the way, and I vowed never to run that race again.

I just didn't like going hard and losing big. And so, after an incredibly average season, I just decided to drop it. I wasn't good, so I just wouldn't do it. I felt like it wasn't worth my time.

*I sense that Simon hated not being great and that he was embarrassed not to be.*

The final embarrassing moment that had me leave team sports forever was when my brother was subbed for me in a basketball game. Okay, we all have to come out sometime, but I never went back in simply because they thought my *younger* brother was the better option. I simply couldn't live with that. He has never forgotten about it and he reminds me about it all the time. I think all of this just fueled my hunger to find my purpose, my passion.

*Okay, now I'm convinced. Simon Cowell hates being embarrassed more than he hates losing.*

### FINAL TAKE: SO, SIMON, IN CONCLUSION, YOUR KIDS WON'T PLAY SPORTS?

Wrong! They absolutely will. The idea that this generation sits in front of a computer, with their gray complexion, silently working their console instead of interacting, playing, and competing, is something I just would not permit.

My message is simple: Play—win or lose—but play. Know your limitations, and in this way, you'll learn about yourself. Most of all, sports teaches communication and teamwork.

Find me a TV show without good teamwork and I will show you a bad show.

### MY WRAP

*Simon is both brutally honest and, at the same time, disarming with his candor. Many who take his verbal barbs would be heartened to hear that he can be just as cutting with himself. He knew he wasn't a great athlete; nevertheless, he tried it, bailed out, and then tried something else. He was honest with himself and has no regrets. He grew to know what he was good at and is not shy about stating it. In a way, his sports frustration has led to his stunning show business success.*

# JEFF IMMELT

★ NAMED ONE OF THE WORLD'S BEST CEOS FOR 2005–2006 BY *BARRON'S* MAGAZINE

★ CEO OF GENERAL ELECTRIC

Sports do not build character. They reveal it.

—HEYWOOD HALE BROUN,
*writer/sportscaster*

$S$ ports were always a very big part of my life. They helped shape me, and I can't imagine my life without them. From day one, when I started at three years old, I played like it mattered. My dad loved sports, and I think having an older brother who played also contributed to some of the success I had.

In baseball I was always a pitcher; in basketball, I played center. Somehow, it seemed I was always in the middle of one season or starting another.

*Sound like a CEO yet?*

### THE COACH WHO MATTERED

I had a high school basketball coach with whom I'm still friends today. He was a screamer, but always in a positive way. He was a little guy and all of us would tower over him. But he earned our respect and attention because of the way he carried himself. He ran our asses off, and I was both afraid of and motivated by him. He had a vision of what he wanted us to be as a team and as individual players. I was captain because I could match his intensity and would question him. I was a little ballsy and a pretty good player. I also knew to stand up to him in private and not in front of the whole team, which made him feel comfortable with me. He liked that and he learned to use me as his translator, a bridge between him and the rest of the team.

### HOW IT HELPS TODAY

Playing on that team taught me a lot about being a leader. I learned that in order for me to be successful, my teammates had to be successful. I also learned how to adapt to the coach's style and be successful, which has helped me today.

### REALITY STRIKES

*Jeff is incredibly humble, and it's clear that he was having fun playing ball while having a fair amount of success, especially in football. But go pro? Not after one particular game in the midst of his best season.*

I always had an idea I would play in the NFL. My senior year of high school, I played in the East-West All-Star game. I went up against a guy named Eddie Beamus, who was on his way to Ohio State on scholarship. He was kicking my butt all over the place. He just annihilated me. I couldn't believe how he crushed me every play, all throughout the game. That's when I knew that I'd better find something else to do with my life. Fortunately, my grades, along with football, got me into Dartmouth.

## DARTMOUTH

I played the line in college and was captain of the team for all four years. It's a position that gets none of the glory and all of the scrutiny from coaches. Think about it. You're on camera every play of every game, and when you watch the game films all you can see is how you did every play of every game. You can't see the fullback or the linebacker, because every game film starts at the line of scrimmage. We would get graded every game and it was all up there for anyone to see. Despite the fact that blocking, hitting the sleds all day every day, is boring at times, it was great. We developed a sense of humor about it, which you really have to have as a lineman in order to survive.

## HOW IT HELPS NOW

In business, just like in my days as an offensive lineman, I always feel like I'm under scrutiny. But I'm okay with people watching me, and I am more than willing to be accountable for what I do. I'm not a guy who needs a lot of credit, which I think is the case with many athletes.

*Where does the personal drive come from?*

These days, when my father says to me, "I wish you would just slow down. Why do you have to be so competitive?" I have to laugh. I can remember coming home from a baseball game when I pitched a two-hitter. My father said, "That's great, but some other kid is out there throwing this afternoon to get better. No matter how well you do, you can do better." So when he says that, I just say to him, "Dad, you created all this."

You can't separate sports from who I am.

*Who would want to?*

**MY WRAP**

*What I get from Jeff's story is that it's all about the effort. If you can handle football at an Ivy League school, you may not wind up in the NFL, but you might just find yourself ready to steamroll your way through the business world and not have to ice down at the end of the day.*

# DAWN STALEY

★ **HEAD COACH, WOMEN'S BASKETBALL, TEMPLE UNIVERSITY, 1999–PRESENT**

★ **USA BASKETBALL FEMALE ATHLETE OF THE YEAR, 2004**

★ **OLYMPIC GOLD MEDAL, WOMEN'S BASKETBALL, 1996, 2000, 2004**

★ **USA BASKETBALL FEMALE PLAYER OF THE YEAR, 1994**

The important thing in the Olympic Games is not to win but to take part; the important thing in life is not the triumph but the struggle.

—BARON PIERRE DE COUBERTIN,
*modern Olympic games founder*

## THE CHALLENGE

I was never chosen for the first game in the playground. No matter who picked the teams, I was not perceived to be good enough and it was a source of great early motivation for me. Maybe it was because I was a girl or thought to be too small. But whatever the reason, I was left out and seething.

## CONFLICT

Tara VanDerveer with USA Basketball and I just combusted. I was on the national "B" team and she challenged me. She asked if I thought I would be called up if their starting point guard on the "A" team broke her leg. I said yes, and she fired back in front of the whole team, "No! You turn over the ball too much." It would not have been as bad had she not done this dressing-down in front of the whole team,. I accepted it because it wasn't personal, it was about basketball, and I knew I could be a better ball handler. I came back the next year a better ball handler, just to prove her wrong. The result was that I was a better player and was wrong to want to prove anything to her. She saw I could take that type of challenge and threw it out there. She was on my side all along. What happened to me then is something I take with me today as I coach at Temple.

## HIGH POINT

My first gold medal was the most important feat I've accomplished. It was just a dream come true. I was coming off a great college season and thought I could hold on for the title on the world stage. It also helped that we were in front of the home crowd in Atlanta.

## RESULTS

Too many times I get people who are so into the results that they forget about effort. It's effort that will get you where you have to be. I always look for the hustler over the natural talent. Now, if I can get a player who combines both, I have a true superstar and a leader.

## LET'S HEAR IT FOR THE BOYS

Playing with the guys helped me with my quickness, my toughness, and my aggressiveness; few were ready for it when I brought my game to the girls' side. My mistake was not having patience with the girls when I saw their game was lacking. Now I push effort over perfection, creativity, or anything else. Give me a player who hustles or a teammate who's trying and I guarantee I'll get results.

## WHY SPORTS IS MY LIFE

Sports to me are the last place we can count on a level playing field. The work you put in equals what you'll get out. There is no knowing the right person, making the key phone call, or paving the way because of someone else's influence. The message I send to my players is: put your life and destiny into your own hands.

## MY WRAP

*Dawn is the standard for American point guards. In terms of toughness and leadership, there is no one better in the world. She never asked for any favors and almost everyone who's seen her play demanded more, which helped make her great. Like soccer stars Julie Foudy and Brandi Chastain, Dawn learned toughness playing against the boys. So few leagues provide that opportunity today.*

# AUGIE GARRIDO

★ UNIVERSITY OF TEXAS BASEBALL COACH 1997–PRESENT; FORMER COACH, CALIFORNIA STATE, FULLERTON

★ WINNINGEST COACH IN NCAA BASEBALL HISTORY: 36 SEASONS, 1,600 VICTORIES

★ 5-TIME NATIONAL TITLE WINNER—1979, 1984, 1995, 2002, 2005

★ 5-TIME COACH OF THE YEAR—1975, 1979, 1984, 1985, 2002

Managing is like holding a dove in your hand. Squeeze too hard and you kill it; not hard enough and it flies away.

—TOMMY LASORDA,
*former manager, Los Angeles Dodgers*

Baseball is not kind. If you're not careful, it reveals all your weaknesses. I won't pay homage to it, but I respect it and enjoy trying to survive and thrive in it. I would rather not have a job than to be forced to teach my kids only baseball. I use baseball to teach about life. Sports creates a metaphor for life and creates the opportunity for people to look inside themselves and solve the crises that they're going through at the time. Once we overcome those crises, we make changes, usually for the better. That's the value of the game, from my point of view. I am not in the program because I'm about money. This game plays on your self-esteem. It challenges you. And that's why I love it.

### AUGIE THE PLAYER

When I was in my fourth year playing in Triple-A, I knew I didn't have it to be a star in the big leagues, so I made my decision to start my coaching career. I figured I was going to end up coaching anyway, so why not start now?

*What? Give up a major league dream just as you approach your window of opportunity?*

I was focused on not doing what my dad did for a living and that was working in a shipyard, so I had to go to college in order to become a coach. Which, in retrospect, is the life he wanted but couldn't have.

### AUGIE ON WINNING AND HAVING GOALS

When you remove the prize and feel your joy, peace, motivation, and happiness inherent in the work ethic, then and only then will your true brilliance emerge. Goals have to be for you, not anyone else. You also have to determine between short-term, intermediate, and long-term goals. Although we usually contend for a national title, I never talk to my team about winning. I talk about the national championship game all the time, but I say that if we get the three outs on defense, that if we throw to the right base, then that's the process. Winning a game is the intermediate goal; winning the national championship is the long-range goal.

## GOT THE PEACE

Watch me during a game and you'll see that I'm not filled with stress. I'm happy because I have peace, joy, and all the things that are important to me. I am not well-balanced. I only care about my team, my family, and my friends. I do not have any hobbies. I do not need diversions in life. I don't need vacations. It's about doing things that matter, like being philanthropic, like giving of yourself. I am not going to be defined by whether my team wins or loses a game. If I lose, it's an opportunity for me and my players to go out and try to fix what went wrong.

## PRESSURE: "BREAK IT DOWN"

I tell my players to remove the prize and that will remove the pressure. Let's say the bases are loaded with one out and a man on third. Does my player feel pressure to hit a sac fly or a single? Those things are the goal. The result, I tell them, is the hitting routine. Forget the circumstances. It's all about their hitting routine. As they go into the batter's box, they should ask themselves about their weight distribution, the feel of their hands on the bat. They should line up their knuckles. They should know that they have the plate covered. They should let go of thinking about anything else. As a result, if their mind is in the right place, there should be no room to feel any pressure. If they think, "What if I strike out?" or "What if I miss the pitch?" then they're driven by fear and they're doomed to fail.

## WHY I STAYED IN COLLEGE AND DIDN'T GO TO THE MAJORS

In professional baseball, the people are used for the betterment of the game. In college, the game is used for the betterment of the people. I am about people and this is why I belong in the college game. My job is about relationships, and I would hate to give that up.

**HOW DO I PLAY THE GAME?**

I tell my players that there is a goal that's higher than winning or losing the game. It's about them and what they learn *after* winning or losing the game. That's more important to me than the actual result.

**MY WRAP**

*How many of you would want to be at a place in life where Augie is right now? He radiates kindness, and his credo blows up in the face of anyone who says winning is everything.*

# DANNY WUERFFEL

★ NFL QUARTERBACK, NEW ORLEANS SAINTS, GREEN BAY PACKERS,
CHICAGO BEARS, AND WASHINGTON REDSKINS, 1997–2002

★ 1996 HEISMAN TROPHY WINNER

★ 1996 JOHNNY UNITAS AWARD WINNER

For every pass I ever caught in a game, I caught a thousand in
practice.

—DON HUTSON,
*Hall of Fame wide receiver, Green Bay Packers*

I may be good at keeping my emotions in check today, but in my years in college and playing pro ball it wasn't always like that. I first remember losing it and paying the price in fourth grade at a racquetball tournament. It was a round-robin competition and I was doing well until the final match. I faced this other really good player and beat him in the first game, but in the second game there was a bad call and I got so bent out of shape that I lost 15–4. But I regrouped in the third for the win. As it turned out, they counted total points, not wins, in a tiebreaker, so he beat me 51–50. I never forgot that the reason I lost that tournament was because I let a bad call psych me out. I learned to shake off a bad play or interception rather easily. Why? Because of that racquetball match!

## ANOTHER MOMENT YOU LIKELY DID NOT MISS

My Heisman season, my team—the University of Florida Gators— played Florida State twice. The first time we came in as the number-one team in the country, and I was knocked down thirty-two times and we lost big. I never lost my composure and never stopped getting back up. Through some strange scenario, we had another shot at Florida State, with the winner taking the national title. This time, we beat them 52–20. Being at the top and getting hammered only to later get back up and re-verse fortunes is what I am really proud of. You might think the loss would be a negative experience for me, but it wasn't. I had many people tell me they saw that game and afterward voted me the Heisman because I kept it together despite the beating we took.

## FLORIDA AT TENNESSEE

First drive, fourth and ten, on the Tennessee thirty-five-yard line, in the pouring rain, Coach Spurrier calls a timeout, which means we're going for it. Can you imagine the pressure in front of 107,000 hostile fans? Even Coach Spurrier's eyes were jumping around from the stress of the situation. Well, out of nowhere, I remembered the C.S. Lewis quote from my bible study class: "As you grow in understanding this concept of faith you will be realized in the midst of chaos. As the world around you is spinning, you will be at the most peaceful place of all." At that point, any stress I had just left my body, because I realized it was only a football game. I went out there and completed a thirty-five-yard touchdown pass, and we went on to win the game and I got the Heisman Trophy.

*We all strive to get to that place on the field and in life: when stress goes up, you calm down. Why is it that Danny is able to make it work? Read on for the answer.*

## ATTENTION, PARENTS

When I was in first grade, I remember racing for ribbons. I ran the 100-yard dash and won. All week I heard my voice in my head saying, "Danny, you will win. You are the fastest and you will come out on top." I remember taking a test, not knowing an answer but knowing that the girl next to me did have it right. And as I went to look at her paper, a voice in my head said, "Danny, you do not need to do that. You're a smart kid. Just do your own work." I didn't cheat, at least not on that day!

Last story. Senior year, I was 6'2", a smallish center on my high school basketball team. For the title game, I was matched up against a guy who was 6'6" and about three hundred pounds. For four days I heard this voice in my head saying, "Danny, you are so strong. No one is stronger then you. You can handle this guy." I ended up doing very well, and our team ended up winning. I never thought about that voice until I had my son, Jonah. My mom came to visit and one day I was getting up early to get a drink at about 6:00 AM and I heard her saying, "Jonah, you are such a good boy. You are so strong and smart." I checked back and thirty minutes later she was still saying it, and it just brought me to tears. I realized those voices I'd heard throughout my life were the voices of my mom and dad, who had been preaching affirmations to me since I was an infant.

## PROVING YOURSELF

I knew I could relax because I knew I was loved and respected through my faith, both before and after every game, win or lose. If your identity and self-worth are tied up in your performance on the field, then you are on a slippery slope. A guy who wins it all and still says something is missing fits into that category. I did not want to have a career where my self-esteem was centered on my stats or whether I got cut. It's part of the reason I could do so well with a coach like Steve Spurrier, who likes to point out what you did wrong and often yells it to everyone within earshot. All that never bothered me because of what I know about faith and football.

## MY WRAP

*Like so many other athletic performers, Danny achieved great things when he stopped caring so much. He seems to have always been older than his years, which explains why he left football when he did in 2002 rather than squeeze in a couple more years as a backup. Danny didn't need football, but football needs more people like Danny.*

# LENNOX LEWIS

- ★ ONLY BOXER IN HEAVYWEIGHT HISTORY TO WIN THE HEAVYWEIGHT CHAMPIONSHIP ON 3 SEPARATE OCCASIONS
- ★ CAREER RECORD: 41 WINS, 2 LOSSES, 1 DRAW
- ★ OLYMPIC GOLD MEDAL, BOXING, 1988 (SECOND-ROUND TKO OF FUTURE HEAVYWEIGHT CHAMPION RIDDICK BOWE)

First your legs go. Then you lose your reflexes. Then you lose your friends.

—WILLIE PEP,

*former featherweight champion*

I was a hyperactive kid and if it wasn't for sports, I was headed for deep trouble. And if it wasn't for my first trainer, Arnie Boehm, who told me at fifteen to "come to boxing, you can be a good boxer," I shudder to think what would have happened to me.

I was also playing basketball and football, but having a place to go after finishing those sports, in my case the gym, rather than coming home to an empty house, was key for me. I just loved learning how to box. I loved the discipline it demanded, and more importantly, I knew I could not be fighting on the streets anymore. Before taking up boxing, I was defiant on the street, and I knew those days had to be over.

## NOT A TEAM SPORT GUY

People now think of me as only a boxer, but those who grew up with me in Canada know I played all sports. My two top sports were boxing and football. I started to get this reputation of being a prima donna because it seemed that every time there was a big football game, I had to go away for a big boxing tournament. My teammates started to resent me because I was not there to score touchdowns for them. It was agonizing, because I felt I had a future in both sports if I worked at it. But I had to make a decision.

My senior season our football team was unbeaten, but we blew the championship by losing in the finals. My teammates were crying and I thought to myself, "Why are you crying? We lost because we made too many mistakes and on the last play of the game our receiver dropped the ball." It was at that point that it occurred to me that I'd rather count on myself then fifty other guys, so I left football, left all my friends, and chose boxing. I wanted to be in control of who wins and loses.

## WHO TAUGHT YOU TOUGHNESS?

It was my mom who deserves the credit for teaching me to be tough. I didn't have a father, so she had to be both mom and dad for me. If I ever fell down, she'd be the first one to say, "Get up, it's not that bad. Other kids are much worse off than you are."

### NO FEAR OF NOT WINNING, JUST A DISLIKE FOR GETTING PUNCHED

I don't know if there was one bout that convinced me I could be good at boxing. I just wanted to think of it like a game. My mind-set was, try it and let's see how far I can go. It turned out I created a stir and got quite a lot of attention after a bunch of wins. But what probably convinced me that I could be real good was a fight in 1984 against the best amateur heavyweight in the world, Tyrell Biggs. In round one, he just outboxed me. I was this eighteen-year-old kid and he just took me to school. Right before the second round began, I sat on my stool and decided that he'd have to kill me to win. It was the first time I was willing to get injured to win. Technically, he was better than me, but for the first time my toughness outshined him. Even though I lost the bout, I sent a couple of messages to the boxing world and to myself. Number one, I could be the best, and number two, I had a lot to learn before I could turn pro. I would wait until the 1988 Olympics before cashing in. I've always been more into the glory than the money. Winning the Olympics would bring the glory and the gold, and later on I'd get the belts and the bucks.

### SURE I GOT KNOCKED DOWN, BUT I GOT UP

I got knocked down twice in my career, and I'm so proud that both times I got back on my feet. The first time was against Oliver McCall, and the second was against Hasim Raham. Not only did I get up, but I got them back in the ring, avenged the defeat, and got back the title. In between my losses and rematches I made sure that I tried to be a good loser, praising my opponent and hailing his knockout punch, while also declaring in a nice way that I would be back.

### HOW THOSE SPORTS LESSONS HELP HIM TODAY

Even today, I'm tireless when it comes to the projects I have to get done. I go at it like I'm still that twentysomething boxer on the road, in the snow, putting in that roadwork. It's that "whatever it takes" attitude that got me through training for my fights, and now it's what guides me through life as a father, businessman, and broadcaster. I always feel like I have to be the one to run in the rain because the other guy doesn't. That's probably what explains how I won.

**MY WRAP**

*He may have been a sensational athlete and a dominant champ, but even Lennox Lewis suffered defeat. Even though he is now fabulously rich and successful, it was the interest of one man that was needed to get him off the streets. Next time you think you are too busy to coach and marginalize the impact you could have on a kid, think of the trainer who decided to take an interest in a fifteen-year-old kid named Lennox Lewis, who would eventually retire as heavyweight champ of the world.*

# STEDMAN GRAHAM

- ★ CHAIRMAN AND CEO, S. GRAHAM & ASSOCIATES, AN EDUCATION COMPANY
- ★ BUSINESSMAN; MOTIVATIONAL SPEAKER
- ★ AUTHOR OF *YOU CAN MAKE IT HAPPEN EVERY DAY*
- ★ 1,000-POINT SCORER, MIDDLE TOWNSHIP HIGH SCHOOL BASKETBALL TEAM

You don't know what pressure is until you play for five bucks with only two in your pocket.

—LEE TREVINO,
*PGA golfer*

For me, sports was a way to feel better about myself because in the environment I grew up in, the color of my skin, being labeled as not good enough, it was one way to showcase my skills without having to worry about race.

I lived in Whitesboro, New Jersey, an all-black town in a mostly white county, and people used to say nothing good ever came out of my town, so we all had to prove ourselves. We had to be twice as good as anyone else, and the way to be revered was to play sports and then shine while playing them. In my case, some people in the black community thought I was "too white," so I got it both ways. But instead of beating up people who insulted me, I figured the way to get back at them, the way to get your name in the paper so people would know who you were, was to get known as an all-state, all-county player. Only then would your confidence be raised, and you'd have a good foundation on which to build other parts of your life.

## MENTOR NEEDED

I played basketball every day of my life, but I didn't really understand the process of being good at the game. What I needed to do was shoot five hundred baskets a day and work on my legs, all while learning to do the simple things right. When I got to college, my coach was astounded by my lack of technique. He couldn't believe that no one had ever taught me the right way to shoot. He made me break down my shooting motion and then rebuild it.

## ATTITUDE ADJUSTMENT

I can only describe my attitude on the court for much of my youth as being much like Terrell Owens today. I was jumping all over my teammates, telling them what they were doing wrong, telling them what they should be doing. I was out of control. In one game in particular, it all came to a head. It was a championship game against Wildwood Catholic and I was just not getting the ball. Throughout the game, I was yelling at my teammates to get it to me. After the game, a guy I respected pulled me over and said, "You embarrassed me and our school and made the entire team feel bad." It was like being hit over the head with a bat, because I really was not that kind of a person in real life, and the thought of showing up my teammates in front of so many people is something I feel bad about today. Looking back, it was me trying to overcompensate for those who didn't think I was good enough, for my skin color, and for having two learning-disabled brothers. It was an example of my low self-esteem and lack of confidence. I thought I didn't belong out there.

## WHAT IT TAUGHT ME

That incident taught me to keep my mouth closed and do the best I could. Just because I might have been the best player on my team didn't give me a license to make others feel bad about their game or themselves. I hated that it happened, but I might have needed it to happen because it helped me survive corporate America and it taught me how to build relationships with people. It's hard for me to believe I could have learned this anywhere but on the basketball court. Today, I know how to handle the ego needs of others by identifying what I needed growing up. I didn't believe in myself, but sports gave me the focus I needed. Sports also taught me to show up, work hard, and be part of a team. Those attributes alone gave me so much more than most of my family and almost all of my friends.

## WHAT DID I HAVE?

I had a mom who would not be slowed down despite our circumstances, and despite the challenges my brothers' situation presented for her. I saw my dad work hard to keep our family and his business afloat. I also had a community that went out of its way to make sure I knew they believed in me. Most of all, they would not let me quit, and to this day I will never quit.

## OPRAH WINFREY

My relationship with Oprah works because we have separate success. Some say, look at what she has. Well, I do look at what she has and I think that's great, but it's hers, not mine. I have had my own path. I've written ten books; I speak sixty times a year around the world. I have my own company. Oprah is not doing that. I have had to work for everything I've ever gotten, and none of it has anything to do with Oprah. No one ever did anything for her, either. She's been doing this since she was seventeen. She doesn't just show up on TV every day and get a script. This is her life.

## MY WRAP

*Stedman had so many excuses to not be successful that few would have blamed him had he not panned out as a star basketball player and later as a star in business. Most of his old friends are either dead or allowed drugs or alcohol to destroy them. Instead of following their lead, he dealt with his painful lessons, broke out, and went back to his roots. What we can learn from this is that your ugliest moments might just be your best self-teaching tools.*

# STEVE YOUNG

★ **NFL HALL OF FAME, 2005**

★ **NFL QUARTERBACK, TAMPA BAY BUCCANEERS, SAN FRANCISCO 49ERS, 1985–1999**

★ **7-TIME PRO BOWL SELECTION**

★ **SUPER BOWL CHAMPION, SUPER BOWL MVP, 1995**

★ **NFL MVP 1992, 1994**

You have to know when and how to go down. The key is to have a fervent desire to be in on the next play.

—JIM ZORN,
*former NFL quarterback, on the art of scrambling*

My dad was the most influential person in my life because he taught me about commitment. He showed us that when we played a game, we had to take it seriously. We didn't miss practice. We didn't quit in the middle of a game or the middle of the year. We did not miss a single practice, a single game. Whether it was basketball, baseball, or hockey, I learned the right way to play.

In 1991 I started my twenty-fourth season as a football player and only then did I think, "I'm pretty good at this.

*Like me, you must be thinking that Steve's being sarcastic. But is he?*

I never was someone who thought I was that good, and I was always trying to prove to myself I could play.

*I guess he's telling us the truth.*

## WORKING IT OUT AND NOT CALLING IT QUITS

My first really tough time in sports came in baseball. As a freshman in high school, I went an entire season without getting a hit. Not one hit. I think I ended up 0 for 46. I was embarrassed and I just wanted to hide. It was one of the first times I can remember having trouble in sports and consciously focusing on getting better, solving the problem rather than just quitting. When the season ended, my dad and I worked all summer to get my game together. I didn't struggle again all throughout my last three years of high school.

## ALWAYS KNEW THERE WAS MORE TO LIFE THAN SPORTS

If you know about my spiritual background *(great-great grandson of Brigham Young, former president of the Church of Jesus Christ of Latter-day Saints),* then you'd know that even as a kid, sports was not going to define my life. I was always worried that I didn't deserve the money, or that I wasn't living up to my or the fans' expectations. For a while, I didn't cash my checks during those years backing up Joe Montana because I didn't feel like I was earning it. I never saw myself as a future pro and certainly not as a superstar.

### THE MOMENT

When I was a freshman at BYU, it was my uncle who pulled me over and said, "Steve, you could be a great quarterback." It was the first time I ever heard someone say those words to me. It made me uncomfortable to hear it, so I used to tell him to shut down on the accolades. The next thing I knew, my college career kicked into gear and I became an All-American and runner-up for the Heisman, and was taken in the first round of the NFL draft. I was on my way. Turns out my uncle was right, and he brought it up last year when I was inducted into the Hall of Fame.

### EXPECTATIONS AND DOUBT

When I went to the L.A. Express in the USFL, they announced I was making $40 million. So here I go again, wondering if I was worth that much, and I felt challenged to see if I could play up to that kind of money. There was a pattern of putting myself in situations where there were massive expectations. I mean, I had to replace Jim McMahon at BYU after he set dozens of NCAA records, and then Joe Montana with the 49ers, having rung up four Super Bowls. Of course, I had doubts and felt pressure. But looking back, I loved the challenge. It was awesome. Through it all, I really felt like I was on my own. It was tough, but I think I'm a better person having gotten through it depending only on myself.

### JUDGING ME AGAINST ME

It's rare to have a supertalented athlete who has a super work ethic. In high school, I recall seeing these great athletes with no drive or dedication and I said, "That won't be me." I was on a quest to see how good I was. My quest was not to set records, but to see how good I could be. Although my approach to football worked for me, it was also a curse to have that voice in your head and never be able to turn it off. You only play once a week, and yet all you do is think about the past and look ahead at what's next.

I was a blamer for a while. I was a guy who would say, "The receiver turned the wrong way," or "The sun was in my eyes." But that didn't work, so I had to learn to put the burden on myself. After all, I was the one with the ball in my hands.

## ANOTHER CHALLENGE

I was bored holding a clipboard backing up Joe Montana, and I didn't feel like I was accomplishing anything—I saw my career slipping away—so I went to law school. By the time I graduated, I was named MVP twice.

## SOMETIMES YOU LOSE

If I lost and gave everything I had, I would still have anxiety about my game, but I could live with that. That didn't mean I was satisfied, though. If you compete you will always go up against people who are going to be better than you, and at times they'll beat you. Even if you're the best ever, it doesn't mean you'll be the best ever every time.

## MY WRAP

*Steve is a study on living up to huge expectations. He learned to look at his opponents as challenges rather than fearing failure. There are few people less impressed with Steve Young than Steve Young. He had to replace two legends in Jim McMahon and Joe Montana, and as worried as he was, he rose to the occasion. It's that success and that effort that helped him develop into who he is today, as a person and as a broadcaster.*

# MIA HAMM

★ **NATIONAL SOCCER HALL OF FAME, 2007**

★ **UNITED STATES WOMEN'S NATIONAL SOCCER TEAM, 1988–2004**

★ **2-TIME FIFA WORLD PLAYER OF THE YEAR, 2001, 2002**

★ **OLYMPIC GOLD MEDAL, 1996, 2004; SILVER MEDAL, 2000**

The most intangible aspect of winning and losing is the human heart.

—MIKE REID,

*NFL Pro Bowl defensive end and Grammy-winning songwriter*

I want to be the best player. That's what I told my college coach, Anson Dorrance. His reply stumped me. He asked if I knew what that meant. I didn't. So he walked over to the light switch, flipped it on, and said, "It's a decision, that's all, a decision. Being the best *you* means being the best every single day." Its sounds strange, because I was coming off a couple of national titles at North Carolina and I'd made the national team. I was a starter on the U.S. National team at sixteen, but now, for the first time, I was learning about true dedication and commitment. I told Anson that I was ready, and so he let me know that that meant that every day I had to think of some aspect of my game that I was going to get better at. So I did just that. My junior year, I trained like crazy, especially on fitness.

**THE MOMENT**

One day, as I was running sprints in the park, Anson happened to be driving by. When he saw me, he stopped and just stared at me. He may have wanted to say something, but I was breathing so hard, he left. The next week, I got a letter in the mail saying, "The vision of a champion is someone drenched in sweat, bent over from exhaustion, when no one else is watching." I still have that note to this day.

At that moment, I felt I was finally on my way. When no one is there to make sure you are accountable, do you push yourself as hard? The answer is, if you want to be the best, yes. I learned to push myself just as hard in or out of the spotlight. My teammates saw the change in me, and as a result, I felt like I was more equipped to lead.

Throughout the rest of my career, I built off that and, in essence, became my own coach. I did not need someone to push me. I knew what to do and how to do it. Myself. No doubt, some of my teammates were intimidated by my intensity, but they also knew I would do anything for them. If I could make them feel good about themselves by making them feel successful on the field, then I would do anything to make that happen.

**THE PUNCH**

Growing up, I played on an all-boys team and it was great because it toughened me up. One day I scored a goal and this kid just refused to admit it was a goal and would not play until I agreed with him. I wouldn't budge, so he took his ball and was walking away when he shoved me. I shoved him back, so he slugged me right in the face. My nose just started gushing blood, but I still wanted to beat the crap out of him. My coach held me back and later he let me know that the way to beat someone like that was to beat him on the field. And the next day, that's exactly what I did. It was one of the first times I ever stood up for myself. From then on, I've never backed off, but at the same time, I've never put myself ahead of the team. If I'd punched him on the field, that would have been putting myself ahead of the team.

I often wonder where that guy is today.

**MY WRAP**

*Mia Hamm had to learn to be tough and to maximize her talent, which made her a leader. It's refreshing to hear that the most naturally gifted female athlete of this generation had to go through the same struggles most of us do. Nothing, absolutely nothing, comes without hard work.*

# MICK FOLEY (MANKIND)

★ **BESTSELLING AUTHOR OF** *HAVE A NICE DAY: A TALE OF BLOOD AND SWEATSOCKS*

★ **WWE/WCW WRESTLER, 1985–2006**

There's such a thin line between winning and losing. Yet the laurels only go to the winner. The rush is always to the champion.

—JOHN R. TUNIS,
*juvenile sports novelist*

M y dad made it clear we had to participate in sports, but winning and losing were never a priority in my house. He wanted us to be part of a team, and he made it clear that I was to play any three sports. But he also said that if after sophomore year in high school we didn't want to play anymore, we didn't have to. I ended up playing lacrosse as a junior, and I also fell in love with wrestling. At the time, I had no idea how much wrestling I would do of a different kind for a career.

## THE MOMENT

One day my friend John McNulty walked up to me and said wrestling would be a good sport to take to get in shape for lacrosse season, so I signed up. It so happened I knew the sport well because my dad had taken me to matches since I was a little kid. I used to wrestle my brother, who had already discovered the sport. I was a lot bigger than him, but I ended up acquiring a little guy's style—which meant being a leg wrestler.

I had some success and was able to work out every day with Kevin James, from *The King of Queens*. He was a starter on the team until he hurt his back, and then I ended up moving in as the heavyweight.

## MOVE TO PRO WRESTLING BRINGS DAD AND SON TO CENTER MAT

My dad and I liked to watch the matches, even though we knew it was entertainment, but our relationship became strained as I got older, trying to live up to the legacy of Dr. Jack Foley. The only thing we didn't argue about, in fact agreed on, was our interest in professional wrestling. When I went into it, few, including my parents, thought I would be successful, but I did it anyway. At first I wasn't crazy about it, but I grew to love it and came to think of it as a great blend of theatrics and athletics.

## THE ACCIDENT THAT PAID OFF: PLAYING WITH PAIN

I got my ear nearly cut off; I had six concussions and a broken jaw; broke my nose twice; separated my shoulder; had fifty-four stitches; a fractured shoulder; and second-degree burns. All these injuries happened in the ring, but the injury that happened out of the ring was my big break. I got into a car accident and lost my two front teeth and truly thought my career was over. The era of the missing tooth-wrestler had passed. But it turned out not to be a problem because a producer saw me and loved it, and my bank account grew.

What I take most pride in is that I never missed a match because of an injury. I just wish I'd brought that work ethic, playing with pain and working hard all the time, to the high school football and lacrosse field. I know what I've learned in wrestling helped me become a better man. In the end, it brought my dad and me back together, and that was enough of a reward for me.

### SETTING GOALS, THE MICK WAY

When I started, my goal was just to get in one match. But you raise your standards and I raised mine to get better, then to join the WWE, then to wrestle at Madison Square Garden. I did all that and more, but I never thought too far ahead. I just played as hard and for as long as I could and never worried about winning or losing.

### HIS REP AND WHO DESERVES THE RAP

I was known as one of the nicest, hardest-working, play-in-pain guys in wrestling history. True, the matches are scripted, but the hits, the falls, and the injuries are far from fake. I got that the work ethic, that ability to absorb pain, from my dad. My dad was my high school athletic director and revered in his profession. He was the hardest-working guy I have ever known. It was not easy growing up where I did with a name like Foley, simply because of his reputation.

### HIS MESSAGE TO KIDS

My message to kids of all ages is to make your own standards of success. I tell them that I have met the rich, famous, attractive, and powerful, and I have always found myself most honored to be around people who have chosen to dedicate their lives to helping others. If it was not for wrestling, I would not have had the fame to bring that message to the kids or to make an impact on any level.

**MY WRAP**

*He may not be in one of the traditional sports, but you can learn a lot from watching Mick Foley. His charm and work ethic should be applauded, and like so many others, he had some hard times at home before achieving success. The bar was set high for Mick and he cleared it. If you have a well-known dad or mom or an accomplished sibling, just know that Mick feels your pain.*

*Bennett is wearing number 74*

# WILLIAM BENNETT

★ **CODIRECTOR OF EMPOWER AMERICA**

★ **SYNDICATED RADIO TALK SHOW HOST,** *BILL BENNETT'S MORNING IN AMERICA*

★ **BESTSELLING AUTHOR OF** *THE BOOK OF VIRTUES: A TREASURY OF GREAT MORAL STORIES*

★ **DIRECTOR OF THE OFFICE OF NATIONAL DRUG CONTROL POLICY, 1989–91**

★ **U. S. SECRETARY OF EDUCATION, 1985–88**

Make the hard ones look easy and the easy ones look hard.

—WALTER HAGEN,
*Hall of Fame golfer*

I started playing football in second grade, and it was a mess. No organization, too many guys on each team, and it was often out of control. I did it because my brother did it, but don't think I was very good at first, because I wasn't.

I went to a very good Catholic high school called the Priory and I left it after my freshman year because they didn't have a football team. Instead, I went to Gonzaga, a sports power, and got a chance to play football. I was able to crack the lineup and I had moderate success on a talent-packed team.

## MY POSITION

In America, we like to say that you can be anything you want, but that's not true in football. No matter what I went out for, I always ended up being the tackle. And if I went out for the basketball team the coach would say, "Sorry, you are a tackle." I had short legs and a large torso. Let's face it, I was a tackle. I won the position my senior year and held it. It was painful to sit on the bench for two years, yet I was prideful that I was a part of this outstanding program. I played at Williams College and was committed to practicing two and half hours a day, even through some back injuries, and it was a great experience.

## IMPACT PLAYER

When I was secretary of education, I started a program that honored teachers who had the most profound effect on your life. I selected my high school football coach, Mike Warner, a marine who taught me toughness without callousness.

Boys growing up today think toughness is being macho, pushing people around, being the big shot, but in reality it's none of those things. Toughness is the ability to keep going beyond fatigue. It means perseverance. It's a virtue, not a swagger. Mike Warner taught me that. Practice after practice he used to yell out, "You think you're tough, Bennett. Take this. And that." He worked me hard, but I could tell he liked me, so I would respond. This relationship was exactly what my mom had hoped for, because I didn't have a male influence at home and she knew I'd get it in sports.

## WHEN DOES THE FOOTBALL PLAYER COME OUT TODAY?

The last time I felt like it was game time was my debate with Howard Dean. I felt the butterflies before going on, which reminded me of how I would respond right before game time. What finally relaxed me in the game was the first contact I made. Hitting someone settled me down.

*No, he didn't slug Howard Dean. I would have led with that if he had.*

Instead of hitting Howard, I caught him with a verbal jab, a joke. Once I got the laugh, I was ready to go. I get sent out for the big debates and that means taking on Dean and Mario Cuomo, and I am always ready to go. I only want the best—that's why we get matched up with each other.

## WHY PLAY?

Teamwork, self-discipline, and the ability to get control of your passion; this is why I played and this is why my kids played.

## MY WRAP

*Bill Bennett might be the deepest thinker we have in the book, and his insight was tremendous. Like so many others, Bill didn't play for the accolades. He played for the right reasons, and for an education expert to choose a football coach as his MVP says a lot.*

# ARNOLD PALMER

★ **62 PGA TOUR WINS**

★ **6-TIME RYDER CUP PLAYER; 2-TIME RYDER CUP CAPTAIN**

★ **WINNER OF 7 MAJORS: MASTERS, 1958, 1960, 1962, 1964; U.S. OPEN, 1960; AND BRITISH OPEN, 1961, 1962**

★ **U.S. AMATEUR CHAMPION, 1954**

A game in which one endeavors to control a ball with implements ill adapted for the purpose.

—PRESIDENT WOODROW WILSON,

*on golf*

My dad was my best instructor. I was strong and coachable, and after a while I realized that if I listened to him and focused I would be successful. His favorite saying was, "Ninety percent of your success in golf will be determined from your neck up."

## MOMENT

It's a funny thing. The more I practice, the luckier I get.

—ARNOLD PALMER

There are a number of moments in sports that I would consider very valuable in my career. The 1954 amateur championship was probably a turning point for me. At twenty-four, I came from behind to beat forty-three-year old Bob Sweeny in the final round to win the title. From August 1954 to January 1, 1955, I won the amateur, met a gal, got married, and started on the pro tour. That tournament gave me the confidence to get on with my life. After spending three years in the service, I had started working as a manufacturer's representative in Cleveland. After two short months, I started the tour and my career in golf.

My direction and desires were all there in 1953 and I felt like I should have won the Amateur, but I didn't. In 1954, it all just came together, which was a result of a maturation process. Even if I hadn't gone pro, after that year I knew all I had to know about being successful in life and in business.

## EARNING IT

The most rewarding things you do in life are often the ones that look like they cannot be done.

—ARNOLD PALMER

My goals were simple: I wanted to go to the Walker Cup and get married. Those were the two things that drove me to do whatever I have in life. The problem was, I could not afford it. So I had to get other jobs to raise the money to do the things I wanted to do. It was difficult, but I think that in order to understand and appreciate things, they have to be earned. If things come too easy then you'll never understand the magnitude of your success or what you have.

## ON WINNING

I never quit trying. I never felt like I didn't have a chance

—ARNOLD PALMER

I was scared of losing, and that made me work harder and harder to accomplish what I wanted to accomplish. That fear is even what drives me today, because I am still not over that fear of losing. I don't care if I'm playing with my wife or my children or a first-rate opponent. I try to do the things I do better than anyone else, and this doesn't just apply to golf.

## THE FIRST IN GOLF AND IN BUSINESS

Concentration comes out of a combination of confidence and hunger.

—ARNOLD PALMER

The differences in my approach in sports and business were stark. In business, I didn't take many chances. In golf, I took chances. I always thought that if I maintained my integrity and knew what I was doing, I'd be successful.

## ARNIE'S ARMY

I hope people reading this will think more about honesty and integrity than championships. The fact that people think of me as playing hard and charging from behind is a great feeling for me. I was, and still today am, about the effort to succeed, not just about the result. That's what keeps me going. I still test myself. I still exercise. I still eat right. And that's contributed a great deal to the enjoyment I'm getting out of my later years. I think I have a large fan base because they recognize the effort I put in. I also try to bring an appreciation of their support every time I come to the course.

**MY WRAP**

*Arnold Palmer is a "whatever-it-takes" guy who maxed out his ability. Few would say he left anything behind in the locker room. As much as it pains him to see his game suffer as he gets older, he'd rather play and record a high score than not play at all. And I think we can all learn from that.*

*Balboa, left*

# MARCELO BALBOA

★ SOCCER ANNOUNCER, ESPN, ABC

★ U.S. SOCCER HALL OF FAME, 2005

★ FORMER NATIONAL SOCCER TEAM CAPTAIN; APPEARED IN WORLD CUP TOURNAMENTS, 1990, 1994, 1998

★ 2-TIME U.S. SOCCER ATHLETE OF THE YEAR, 1992, 1994

If they can't put up with my pressure, how are they ever going to stand the pressure from 60,000 people?

—VINCE LOMBARDI,
*legendary Green Bay Packers coach*

I was a reckless, collision-oriented player as a kid. My brother played very skillfully, like my father, but I loved contact. My dad was a pro player and didn't try to change my style, but he did make me work harder to add more dimensions to my game. He would come home from working nights and go in the backyard and train the both of us. He knew I was a natural defender but always said, "You can't just be a defender, you have to be dangerous and move forward so that you're able to score when the opportunity presents itself."

### DECISION TIME

As a kid, I played every sport, but when I reached fourteen, my dad made me make a choice: baseball, football or soccer? I chose soccer, but he wasn't done. He handed me and my brother paper and pen and said, "Write down your goals in the sport." I wrote that I wanted to make the Olympic team and the national team. From that point on, he'd use my goal sheet to reinforce for me what I said I wanted. It gave me focus and helped light a fire under me to succeed.

### DAD

My whole life my dad was always critiquing my game. He said good things, but he also let me know what I needed to work on. Up to the last game I played as a pro he was studying my game and letting me know how I could play better.

As kids, my brother and I were able to recognize three speech patterns in my dad. We knew what was coming after every game. Number one was a good one—"You did okay." Number two was, "Your effort was barely passable." Number three was, "You suck!" Something must have worked, because when he coached me and my brother, we won most of our games as well as an under-19 (U19) National Club championship.

### THE MOMENT

At fifteen, my dad and I had some problems. He thought I wasn't playing hard enough and I thought he was being unreasonable. After all, I was already one of the best players on the team with or without a monster effort, so why break a sweat? One day, after repeatedly threatening me, my dad kicked me off the team. I was his best player and I thought I'd get special treatment, but that wasn't the case. The team had traveled a lot, and we were just coming off a trip to Germany, where a club team watched me play and offered me a contract to stay there and go pro. Not many kids could handle this type of acclaim, and you can count me among them. So what did my dad do with me and my new attitude? He kicked me off the team and said, "Go play for someone else." At first I thought it might be a good move, but then I realized I didn't have a ride to games and I couldn't even get a tryout without my dad, so I asked him to take me back. He offered me a *tryout* to get back on his team.

I passed and, at fifteen years old, I learned a great lesson that helped me spend years in the starting lineup on the national team: the easy part is getting there; the hard part is staying there. If you're good, you'll get a tryout and maybe play well enough to make the team. But will you outwork all comers to stay there? That's the bigger question. After getting cut, I pledged never to be outworked again, and I don't think I was throughout the rest of my career.

## CONSISTENCY IS THE KEY

My dad was not about having a great game followed by a bad game. He wanted consistency in his players and their game. Why play a great eighty-nine minutes and then have a mental breakdown, let your man break free, and watch him get the game-winner in the ninetieth minute? I needed to be consistent, to sometimes sacrifice being great so that I would never be bad. It was consistency that got me on the national team, but I didn't get a lot of headlines or marketing opportunities, as did Alexi Lalas and Tony Meola. I just had the game and the pride that came from playing it as well and consistently as I could.

## CONFIDENCE LOST

It had never happened before, but thanks to one man's word processor I began to doubt myself. The first year I made the national team, I was the subject of a Paul Gardner column headlined "RAMBO." He wrote that it looked like I was only out to hurt people, that I was devoid of skill, and that I would soon self-destruct. I was new to the national team, and I was crushed. It rattled me so much that I started to think that I really had no skill, so the next game I played I overcompensated and tried to play a finesse game instead of the way I normally played. It was a disaster. Thankfully, my dad straightened me out. He let me know I could play and that there was nothing wrong with my style. Looking back, I realize how fragile my confidence and my teammates' confidence really were. The press didn't know me, but they had the power to destroy me as a player. From then on, I played angry, always in control, but always playing hard from whistle to whistle.

## FINAL THOUGHTS

I just wanted to play well enough to leave a legacy so that one day my kids could look at my career and say, "My dad was really good." And that would be fine with me.

## MY WRAP

*You know the bigger names in world soccer—Beckham, Ronaldo, Landon Donovan—but only the true soccer fan knows how special this U.S. Soccer Hall of Famer was as a player and as a leader. I'm talking about his work ethic, his drive, and his fearlessness. As much as you'd be tempted to focus on the national team coaches who brought him along as a pro, I think those days working alone with his dad are what laid the groundwork for his success.*

*Nixon is wearing number 12*

# RICHARD NIXON

★ **37TH PRESIDENT OF THE UNITED STATES, 1969–1974**

★ **VICE PRESIDENT, 1953–61**

★ **U.S. SENATOR (CA), 1951–53**

★ **U.S. CONGRESSMAN, 1947–51**

The single-mindedness necessary to fight one's way to the top, in no matter what spot, is something not shared by the majority of mortals.

—PAUL GALLICO,

*sportswriter, novelist, author of* The Poseidon Adventure

When I interviewed President Nixon for my book, I quickly picked up on his passion for football. He wanted to play football in high school, but he was prevented from playing by doctors who said he had a tuberculosis shadow on his lungs. It pained him to miss out on the game he loved, because he wanted so much to be a part of the team.

He finally got cleared to play at college. There was a problem, though. At one hundred fifty-five pounds he was too small to play the line, and he was too slow to play running back. His inborn clumsiness made him inept when it came to passing or receiving. He ended up a permanent fixture on the bench, technically not even making the team.

One of the figures who shaped his entire life back in the 1930s was his football coach at Whittier College, Chief Wallace Newman. He taught Nixon to never give up and to keep on fighting. He credited the chief with instilling in him the inspirational dream that by hard work, training, and preparation, even the greatest victory can be achieved. This is not to say that Nixon could play football—he was one of the worst players on the team—but he was still inspired by Newman. I remember one time Nixon told me, "He tried to get all of us to be self-sufficient, to be competitive and . . . never give up."

On the old cliché, "It's not whether you win or lose, but how you play the game," the chief said, "That's nonsense. Of course how you play the game counts, you should always play fair. But it also counts whether you win or lose. You play to win. If you don't win you kick yourself in the butt and make sure you don't make the same mistakes again."

There was story that was relayed to me a few times about Nixon's response when the coach turned around during one game and said, "Nixon, what would you have done in that play?"

"Sir, I would've pulled the blanket up just a little tighter around my shoulders."

Chief Newman was called chief because he had Indian blood and that inspired him. Newman did see in Nixon qualities that helped him stay on the team, and those were tenacity and courage. He may have gotten into only one game, but he never missed a practice, arriving first and staying late. This habit and his approach to the game did not go unnoticed by his teammates, who were constantly amazed by his tenacity.

### TWO TEAMMATES AT WHITTIER

Clint Harris: "I remember thinking, "Dick Nixon, I don't know why you do this, but I admire your red-blooded intestinal fortitude to stay with it until the end of the season.""

Gail Jobe: "Nixon and I were cannon fodder. We were two of the smallest guys on the team, but we learned how to hang in there and smash the big guys back. I'll say this for Nixon, he had plenty of guts when it came to taking a beating, getting up off the floor, and coming back fighting."

Chief Newman: "He was tenacious as the dickens. When he got ahold of something he never let go."

Nixon looked at the chief as an almost mystical figure in his life and he felt okay about being a reserve on the team. That didn't mean he was not embarrassed at not being good enough to be on the field. He hated not playing, but he didn't quit. For four years, he was out there from four to seven PM. Newman was still so important to Nixon that after he resigned from the presidency in August 1974, when he was at the lowest point of his life, it was Chief Newman who said pick yourself up again and fight back. Nixon told me the chief had a lot to do with his reinvention in the later stages of his life.

I don't think Nixon's self-esteem was tied into how he played football. His lifelong yearning was to be one of the boys, and you couldn't be one of the boys and not be part of the football team.

Nixon was a huge sports fan, and I know this because almost every one of my interview sessions was interrupted by or scheduled around some baseball or football game. He was dead serious about his teams and always pointed out to me the character traits of different famous players, character traits that most others would not be able to see.

I really enjoyed watching him watch the games.

**MY WRAP**

*Jonathan went on to say that he almost called the book* Fighting Back *because of Nixon's days on the field and his passion for sports. He was a tough, hard-working guy before football, but there is no doubt that this game and this coach were an integral part of his life. There is also no doubt it helped him battle back from all the trouble he had in his political career, from losing the presidential election to John F. Kennedy, losing the race for governor of California, and through Watergate. He was a fighter who never lost his passion to play and win. He was the first Rudy story, only he didn't get the chance to make the final tackle.*

# PAT SUMMITT

★ **WINNINGEST BASKETBALL COACH (MEN OR WOMEN) IN NCAA HISTORY**

★ **6-TIME NCAA WOMEN'S BASKETBALL COACH OF THE YEAR, 1983, 1987, 1989, 1994, 1998, 2004**

★ **NAISMITH COACH OF THE CENTURY, 2000**

★ **WOMEN'S BASKETBALL HALL OF FAME, 1999; BASKETBALL HALL OF FAME, 2000**

★ **6-TIME NCAA CHAMPION, 1987, 1989, 1991, 1996, 1997, 1998**

★ **OLYMPIC GOLD MEDAL, 1984, COACH OF U.S. WOMEN'S BASKETBALL TEAM**

★ **HEAD COACH, UNIVERSITY OF TENNESSEE WOMEN'S BASKETBALL, 1974–PRESENT**

My idea of discipline is not makin' guys do something, it's getting 'em to do it. There's a difference in bitchin' and coachin'.

—BUM PHILLIPS,

*former NFL head coach*

A couple situations happened that impacted my coaching philosophy and my toughness. The first one was the knee injury I incurred in the fourth game of my senior year at the University of Tennessee-Martin, which ruined my dream of playing on the Olympic team in 1976, after playing in the World University Games in 1973. The injury stopped me in my tracks. I now had to spend my time in rehab, not on the hardwood. It was a defining moment in my life because it played a major role in my future, in terms of becoming a coach. After that injury, which was a heavy blow, I realized how much I loved basketball. In order to satisfy my love for the game, I took the graduate assistant's job at the University of Tennessee. I was able to rehab my knee, but because I was far from 100 percent, it didn't look like I was going to be able to play in the Pan Am games.

### RISING TO THE CHALLENGE

The other player who didn't get to play, because she was the youngest player on the team, was future Hall of Famer Nancy Lieberman. We were playing Cuba and we were twenty-seven points behind with less than two minutes to play and the coach told us to go into the game. Nancy said, "I'm not going in." I guess she was insulted to be playing only because it was garbage time. I told her, "You're going in. Let's go," and we headed to the scorer's table. I think going in helped save her basketball career, because if she had refused to go in, it would have been ugly. Meanwhile, the drama for me personally was just beginning. After seeing me play, the executive director of the team said, "You won't make the Olympic team. The knee is still bad and you need to lose weight. I'm betting against you being able to do it."

I said simply, "You're wrong. I will make it."

### OH, YEAH—I THINK I DO HAVE WHAT IT TAKES!

I worked out hard, stopped feeling sorry for myself, ate healthy for the first time in my life, dropped all the weight, made the team, and was named one of the co-captains by my teammates. Yes, I was an Olympian!

I knew then, and for the rest of my life, that I had the self-discipline to do what it takes in any situation. I later found out that that executive director was pulling for me and threw out that challenge so I would take it and make the team. Let's just say he knew how to offer incentive.

### HOW THIS MADE ME A BETTER COACH

Sitting on the bench gave me compassion for what reserves go through. I had started and stood out throughout my career, and the knee injury showed me the other side of the equation. I also now had a better understanding of the kids who think they have discipline but don't. I thought I had it, but I had to dig deeper and go to a different place. I had to learn how to eat and how to train, and in the meantime I had to sit on the bench and not disrupt the team.

### THE COACH WHO MADE THE DIFFERENCE

The coach who really taught me how to play the game and gave me the mind-set to be successful was my coach at the World University Games, Billie Moore, from UCLA. She had a way of challenging all of us. I had never been through practices so demanding. We'd begin thirty minutes, one-on-one, full-court, and then start drills. She really got a bunch of talented players to become one team, and she did it quickly.

### WINNING?

It makes me physically ill to lose. I don't care if I'm playing checkers or marbles, I want to win. I got that from my father. He would never let us win at anything. We had to learn how to beat him. I grew up as one of five kids and as the second youngest, if my dad wasn't beating me at something, then my brothers were. Also, growing up on a dairy farm, we were working from five in the morning to seven at night. That's where I got the work ethic along with my will to win. My dedication can be found in my school attendance record. My parents didn't let us miss a day of school in all twelve years. I slept through an English class in college once and I was crying to my professor, who told me to calm down because he didn't even take attendance. Principles my dad would also drill into our heads were: don't ever cry when you lose, and don't you dare show anyone you have any weakness in you and never brag about what you have done. If you are as good as you think you are, people will talk about you and you'll get your due. Don't let it come from you.

## HOW DAD HANDLED HIS INCREDIBLY SUCCESSFUL DAUGHTER

My dad gave me so much, but he was a man of few words and he wasn't much for expressing his feelings. In 1996, after winning the national title, he stayed the night with me in Knoxville and he said he loved me. It was special, and although I knew it all along, there's something about hearing it that I will never forget.

## FINAL THOUGHTS

I became a better coach when I went through my toughest time after the knee injury. I never took the game or anything else for granted again. I guess you could say it was the year I grew up. Before anyone says that I've had nothing but success in coaching, remember I went to seven Final Fours before winning one, so I know exactly how hard it is to win at this level. And let me add, I always love trying.

## MY WRAP

*It's rare when a super player becomes a legendary coach, but that's exactly the career Pat Summitt has had. It is heartening, yet noteworthy, that even with her great parents and tremendous work ethic, she was far from a finished product at twenty-one. So for anyone who thinks that they should not stumble during or after college if they want success in life, think of Pat Summitt. For me, it's no longer a secret why she's the best college sports can offer.*

*Robertson, second row, second from left*

# PAT ROBERTSON

★ **FOUNDER OF THE AMERICAN CENTER FOR LAW AND JUSTICE, CHRISTIAN BROADCASTING NETWORK, AND THE CHRISTIAN COALITION**

★ **HOST OF THE 700 CLUB**

★ **AMERICAN CHRISTIAN TELEVANGELIST**

Few lapses of self-control are punished as immediately and severely as loss of temper during a boxing bout.

—KONRAD LORENZ,
*Austrian ethologist (zoologist who studies animal behavior)*

In prep school, we had to pick sports to play and I chose boxing. I didn't do it to be a great fighter, I picked it because I thought it would get me in great shape. I mean, you have the roadwork, the sparring, the bag work, so how could I go wrong? Well, northing went wrong and, in fact, I wound up on the boxing team, competing in the Golden Gloves fighting heavyweights at the age of fifteen. I don't think I have ever done something as intimidating or felt more fear than those moments sitting on a cold locker room floor, waiting to fight a guy who was more than likely bigger, stronger, and more experienced in the fight game. And at most of these tournaments, there were about five thousand mostly hostile fans.

## BIG CHALLENGE, BIG LESSON

Slick Evans was a big, rangy fella, a lot older and a lot bigger than me. He was my sparring partner, and he became like a coach to me. To be candid, I was scared to death of him. So here I am sparring with Slick, keeping my distance, and he offers me great advice. "Pat," he said, "even if you're fighting Joe Louis, don't back off. Go after the guy, because you can't just cover up and play defense in the ring. You have to fight."

And I knew, even at fifteen, that I'd just gotten a great life lesson as well as a boxing lesson.

## THE CHALLENGE

The fighter nicknamed the Ooltewah Terror was a certified hillbilly from the mountains of Tennessee. He was big, angry, and powerful, and I was not looking forward to mixing it up with him. He was my next opponent in the city tournament, so I knew I would not have a choice. Don't you know, I bumped into his victim from the previous day and he explained to me that he'd just gotten out of the hospital after being knocked out by the big heavyweight the day before. This guy had knocked out everyone he had fought in this tournament. Before this guy left me he let me know that the Ooltewah Terror had one weakness: he telegraphed his punches. Well, I get into the ring, the bell goes off, and he races across the ring and nails me with a telegraphed punch that put me through the ropes. I got up, put myself back together, and fought him to a standoff the rest of the way. If I had not been knocked down early, I likely would have won. To this day, I am proud of that, and though I'm seventy-six years old, I still remember the punch and the fight like it was yesterday. Why? Because it taught me to always search for the edge in any situation. Find out what you have that the other guy doesn't have and then try to overmatch your opponent in everything you do. Sure, a man or woman may be stronger, faster, and younger than you, but there is always something that God gives you that could give you an edge, if you choose to use it.

**NO FEAR**

I have run for President and been involved in high-profile debates in front of millions, and I just took it all in stride. Partly that's due to the pressures and fear I felt as a kid in the ring. If I could survive all these big punchers, I just felt I could hold my own in front of all these heavyweight intellects.

**THE WRAP**

*Who would have thought that one of the world's most popular preachers could pack such a wallop and still lift and run today? We all face fear, but how many look to avoid it, and how many climb through the ropes anyway? Pat Robertson may not always win, but he always makes his way through the ropes and into the ring, and, I imagine, always will.*

*Ruettiger is wearing number 45*

# RUDY RUETTIGER

★ MOTIVATIONAL SPEAKER

★ THE MOVIE *RUDY* WAS BASED ON HIS EXPERIENCES ON THE NOTRE DAME FOOTBALL TEAM (HE WAS PLAYED BY SEAN ASTIN)

★ NOTRE DAME CLASS OF 1976

By itself, practice does not make perfect. Those of us with a ten-year-old son practicing the trumpet may understand that.

—DR. DANIEL HANLEY,

*U.S. Olympic team physician*

I got my work ethic from my parents. I was the third of fourteen kids, and we all had jobs to do. I saw my father working three jobs. My mom worked hard twenty-four hours a day to keep the family moving. It's hard to see those examples and be a lazy kid.

What sports did for me was to vent some anger. I was frustrated in school and did not do well. It wasn't until I got to college that I found out I had dyslexia. I was labeled a slow learner, and as a result I didn't feel like I was part of the class. Instead, I applied my work ethic to sports, and that was where I could vent, not just in football and wrestling, but also in baseball. I was the kind of athlete who always hustled. I was the first one there and the last one to leave. I did well in high school sports, because I had the attitude and the drive, but by the time I got to Notre Dame and tried to get on that football team, it was a different story. What I'm saying is that when you get to a program like that, hard work is not enough. I was like, "Wow, what is going on here?"

*Okay, but it also didn't help that Rudy's only 5'6," one hundred sixty-five pounds.*

## CONFUSION

Notre Dame did not want me to be part of the program, and I could not make sense of why my effort was not enough. For the first time, the struggles that I had in school carried over to the football field. I could not understand why effort would not be enough to make my dream come true. I went through a period of disillusionment as I saw kids with more talent, but who were not working nearly as hard as me, get recognized. I thought this could not be more unfair. Through it all, I refused to quit. As a walk-on, you are basically invisible because you're not on the list to go to training and often you don't even have a locker. I often thought I was wasting my time. Little did I know my presence was inspiring the other guys. My work ethic was pushing them to work hard.

## THE PAYOFF

All I wanted was one chance to dress for a game, and I didn't think this was going to be a difficult thing to happen because Notre Dame has a policy that every senior dresses for the last home game and gets a chance to run through the tunnel. But there it was, my last chance, and in comes the NCAA and they ban that practice, so it was clear that I'd have to be on the limited roster. I was still hopeful that my two years of not missing practice, coming in early, and being the last to leave would be enough to include me on the final roster for my final game.

Thursday, they posted the lineup and my name, Ruettiger, wasn't there. I thought for sure Coach Devine would let me dress for a guy who was injured. But it didn't happen. I was crushed and was convinced that this two-year sacrifice and lifelong dream was just a colossal waste of time, that hard work and effort do not pay off.

That Friday, I had three meetings that straightened my head and life out forever, and ruined what I thought was going to be a feel-sorry-for-me-day. (In the movie they made those three guys into one character.) One was with a friend who is a federal judge today. He gave me a new perspective by saying that sports was not an end-all, just a part of life, and because I didn't dress didn't mean that I was a loser or worthless. Instead, it was about being part of the team. The second talk was with a janitor. It was a blue-collar talk essentially saying, "How dare you feel sorry for yourself," and he brought back my own words of how I used to say that just being part of the football program was good enough. He went on to say that "Life isn't fair, so get used to it." He told me how happy he was to be at the university, a place that didn't pay him a lot, but still a place where he was thrilled to work. Last, I came across a player who quit the team because of Dan Devine and he regretted it every single day. He wanted to make sure I did not make the same mistake. He said, "Rudy, I quit. Don't *you* quit!"

As you may know, I got to play in the game. The four team captains went to Coach Devine and asked him to put me on the roster and give me a uniform for my last game. The problem—someone had to offer his uniform up so that I could suit up. Rich Allocco did just that, and I got my chance to see my dream come true. If you saw the movie you know I did dress and I did get in the game. In my only play, I sacked the Georgia Tech quarterback and I was carried off the field. Looking back now, by the time I made the play it didn't matter because I realized I had already won over my teammates with my work ethic and that I'd done the most I could possibly do to be successful, and that was enough.

## IN THE NAVY

I had great parents, but my years in the navy were what set me up for my successful and extremely unlikely journey to Notre Dame. It was in the service that I acquired the confidence to even step onto the Notre Dame campus. My joining the navy set up that tackle at Notre Dame, because it was in the navy where I felt for the first time that hard work was recognized and rewarded. That's what gave me the blind faith to give Fighting Irish football a shot. Of course, there were a couple of problems—I didn't have the money or the grades.

Once out of the service I hatched a plan that included going back to work and getting into prep school at Holy Cross. It was at Holy Cross where I found a way to get my grades up so that I could apply to Notre Dame. After a few tries I got into the university and played so hard at football tryouts that they kept me around. So between the GI Bill and staying in the athletic dorms, it worked out. I was able to overcome the academic and financial challenges.

## YOU NEVER KNOW

Okay, so what do people get from my story? It's certainly not *Try hard and become a football player*. Rather, it is work hard and you never know where it can take you. I never wanted to be a motivational speaker, but through my story and the movie it has become a career for me. When young men and women come up to me and say how my story changed their life, I realize again how important it was not to quit, even though my playing in that final game against Georgia Tech didn't affect the outcome, and my playing on the practice squad for two years had little to no effect on Notre Dame's final record. The only thing you can control is the effort, and you owe it to yourself to persevere and do whatever you decide to do with passion. Don't ask for permission to be successful. Go make it happen.

**MY WRAP**

*Amazing, unimaginable things happen when you don't quit and keep your head screwed on straight. And most of those things are for the better.*

# MYCHAL THOMPSON

★ **LAKERS RADIO COLOR ANALYST**

★ **FORWARD/CENTER WITH THE LOS ANGELES LAKERS, SAN ANTONIO SPURS, AND PORTLAND TRAIL BLAZERS, 1979–1991**

★ **NUMBER-ONE NBA DRAFT PICK, 1978**

Be quick, but never hurry.

—JOHN WOODEN,
*legendary UCLA basketball coach*

I always cared about the game. Even when I was a raw player, I always wanted to look good. On the islands, the Bahamas, I played cricket and soccer and dreamed of being the next Joe Namath. I didn't start playing basketball until I was seventeen. As a kid, when we played we drew a crowd, and I never wanted to be seen as unprepared or wild. I always tried to play my best, even at a young age. I also cared tremendously about how my peers felt about me. I got that from my dad, who was a prideful man who worked hard his whole life. He set a great Christian example for me and my siblings. As I grew out of all those other sports I was pushed toward basketball by my friends and family. Eventually, it brought me off the islands and to the States, where my potential could be reached.

## PRO DREAMS

I knew it was possible to get off the island and into pro sports because others did. I watched Roberto Clemente and Juan Marichal and thought maybe I could do it, too. They gave me hope. I know today kids look at me and see that their dreams can come true because mine did. Because of this, I always go back and give back. When I played, not only did I feel like I was playing for my career and family, but for all of the Bahamas, and this helped me stay focused my whole career. It was a role I cherished. What I am proudest of in my career isn't the titles or the money, but that I was the first Bahamian to play professional sports.

## FEAR OF FAILURE?

Yes, and that's why I prepare and prepare and prepare some more. I still had bad games and today as the cohost of a sports radio talk show, I have bad shows, but at least I know I put myself in the position to be successful. Even if I fail, I just get up and try to be better tomorrow.

## NO STRUTTING—JUST HAPPY TO BE THERE

The way I was able to deal with the fame of being a pro player, a Laker, and a world champion is simple: I was never impressed with myself. I knew I had a glamorous job that I loved, but still, I was no better than anyone else. I always considered myself the island boy, the kid from the Bahamas who shouldn't be there but somehow found a way to slip through.

## THANKS, COACH

Jack Ramsey taught me how to train, how to work, and how to eat. He was a coaching legend and he actually worked out like a player. Just seeing him train and eat was a great example for me and helped extend my career to the age of thirty-six. By the way, I know I could have played into my forties and I regret not giving it a try.

When it comes to winning, Pat Riley taught me so much. I know anyone can win once, but Pat Riley raised my standards. He made me think bigger, and I still take life like that today. I know not to settle, and I also learned how to work and accept having to pay the price.

## HOW I PLAYED THE GAME

I prided myself on intelligence, on trying to outsmart my opponents. I was not the most talented player out there. My work was done before the game: getting to know my opponent. I had to know where their strengths and weaknesses were, and I still do that today now that I cohost a radio talk show. I mean, when you're going against a Karl Malone, a Moses Malone, a Kareem, if you don't do your homework, you'll get killed. Especially because they were all more talented than I was, I had to fill the gap somehow, and for me it was with videotape and hard work. My goal was never to shut anyone out, but just to keep them under their averages.

## HOW IT HELPS TODAY

Today, as a talk show host, I get calls from all around the country, and there are times I get abused verbally. My partner is always surprised that I don't fire back or get excited, or ever take it personally. That mind-set comes from playing and hearing it from the fans at the away games, and even from the hometown supporters when I didn't produce. The more I'm out of sports in real life, the more I realize sports is like life. It's sorrowful. It's playful. And just like life, you're going to lose family members and friends who go their separate ways. Life is like sports because it's so full of ups and downs. There are days when I make great decisions in the same way there were days on the court when I felt like I couldn't be stopped.

## WHAT'S WRONG WITH TODAY'S SPORTS PARENTS?

Too many parents are putting too much pressure on their kids. So many parents look at their kids as a meal ticket and try to live through them, and that's very, very sad.

## MY WRAP

*Mychal gets it. I think that comes from growing up in a neighborhood where the families had very little. Even today, on his sports radio talk show, he says it's a privilege to do something he loves and that, my friends, is talk about sports. How many of you get to do what you love for a living? If you don't, maybe you should try.*

# RUSH LIMBAUGH

★ **RADIO TALK SHOW HOST WITH 13.5 MILLION LISTENERS EACH WEEK**

★ **MARCONI RADIO AWARD FOR SYNDICATED RADIO PERSONALITY OF THE YEAR, 1992, 1995, 2000, 2005**

Sport is the very fiber of all we stand for. It keeps our spirits alive.

—PRESIDENT FRANKLIN DELANO ROOSEVELT

I grew up playing Little League ball until age of ten and then went into the Babe Ruth League, where I had some success. I was a pitcher and when I wasn't pitching I played first base.

I had two equally defining moments in sports that helped me immeasurably in my life and career. The first came when it was time to go out for the high school team. I thought I was a shoo-in. After all, I had played against and with all the great players who were my age and I'd held my own. I went out for the team, but after the first tryout I didn't see my name on the posted list, so I didn't even survive the first cut. I couldn't believe it. I thought it had to be some kind of mistake. What made it worse was that the coach was a family friend, so I thought I had it locked. I was in shock. And when I went home and I told my dad, he couldn't believe it either. So I said, "Dad, do me a favor. Please call the coach and find out what happened." He fired back, "I damn well will not call the coach, son. You didn't make the team and it's time you deal with reality. And if you really need to find out why you didn't make the team, then *you* go talk to the coach." I never called him, though.

You can go one or two ways when something like that happens, because make no mistake about it, I was embarrassed. Number one, blame somebody else or, number two, go through an honest self-appraisal process and learn from it. What was it about me that made me think I was a shoo-in in for that squad when in reality I wasn't?

From that day on, I vowed to examine my failures so I could learn from them. It happens to be one of the most difficult things a person can do, but it's necessary. My decision about my sports career would result in another valuable lesson later on in life. At that moment, I decided to end my quest to play high school baseball and try to become the kicker on our high school football team.

**NUMBER TWO**

The second defining moment came in the fall of 1967, when I walked onto the football field and told the coach that I'd like to be their kicker.

Well, he let me know that they didn't have a roster spot just for a kicker and that I'd have to do something else. He recommended the offensive line. The coach was Norm Dawkins, and he was one of these guys who truly was in the game so he could shape kids' lives. He wanted me to have the respect of the team. He didn't want me to sit around while the team did the "Bull in the Ring" or "Nutcracker" (for you non–football players, these are one-on-one drills made to build character). Essentially, I wanted to be the kicker because I did not want to run, and I think Coach Dawkins knew it. Practices were brutal, one of the most physically taxing things I've ever done.

We ended every practice with gassers (wind sprints) and he'd say after a number of them (during which I'd always be near last), "the top three finishers get to hit the locker room." Well, I had been pacing myself for just this moment and I'd wind up finishing among the top three and be done.

Coach Dawkins called me over and asked how I pulled that stunt off, at which time I let him know I was pacing myself. He ripped into me, saying in this game and in your life you do everything *full-out* all the time, that it's the only way you can reach your full potential. I have never forgotten that. Those words stuck with me, and that principle helped me throughout my broadcast career. If I had gotten away with it, it would have sent the wrong message, that you can fool people, even authority figures, who in reality are only trying to get the best out of you. The coaches in that small Missouri town cared a lot about us. They were intent on shaping young adults, not just football players.

### SO, RUSH, HOW DID THE REST OF THE SEASON GO?

I played every game on JV as a tackle but never got into a varsity game on the line. But I was the team kicker. The highlight for me was kicking the extra point and beating the dreaded Carbondale, Illinois, team. I never did play my senior year, because I found a full-time on-air job at a local radio station.

### SO, WERE ALL THE DRAMA AND DRILLS WORTH IT?

If I'd been cut from the high school baseball team and called out during the first days of high school football, I don't know how I would have handled being fired seven times in my broadcasting career. I never made excuses or looked to blame anyone or feel bad for myself. Instead, I just moved on. Somehow, I never lost my confidence. I even left radio for a few years, joining the Kansas City Royals front office. After five years being around people making tons of money while I was only making eighteen thousand dollars a year, I got out and found my way back to radio news. I convinced my bosses to give me a twice-daily commentary spot. I still got fired, but I had enough of a track record to get a talk show host slot in Sacramento, which led me to where I am today, the most successful radio host ever. But along the way, I had some bad times, including having one ABC network guy tell me, "You don't have any talent, so if you love radio, try sales." But I had a passion for it and was not going to be stopped. I think desire is 80 percent of achievement and I just loved radio, and I think that's why I'm where I am today.

## WHAT I LEARNED FROM WATCHING GEORGE BRETT

The days with the Royals were bleak ones for me, but I did have the chance to watch one of the all-time great baseball players in his prime—George Brett. They say he made everything look so easy and he did, but few knew why it looked so easy. Brett looked effortless because he put the work in all the time. For a night game, he was taking batting practice in one hundred five-degree weather six hours before the game. He did that all the time, and watching that made me redouble my efforts to get back into radio, making me more determined to do whatever it took to get it done.

The latest example I can use from my life is what I went through hosting my radio show as I lost my hearing. I had a syndicator who stood by me and an unbelievable staff working extra-hard for me, but I also did what I had to do to stay on the air. This included my phone screeners typing out what the callers were saying, so I could answer in a somewhat natural way. Sure, I was worried and stressed out, but I worked through it, and it takes me back to my sports experiences as well as my overall philosophy of finding a way to get it done. I've learned to accept reality, and feeling sorry for myself is just not something I want to do.

**MY WRAP**

*Most of us have to bust our butts to get where we are, and it's even harder to stay there. Rush has made it and he's sustained it, and he might just end up back in sportscasting on the side as well. No one could ever accuse him of pacing himself, thanks to a special high school football coach.*

*Kemp, wearing number 49, with teammate Jim Mora*

# JACK KEMP

★ **1996 REPUBLICAN VICE PRESIDENTIAL NOMINEE**

★ **SECRETARY OF HOUSING AND URBAN DEVELOPMENT, 1989–93**

★ **UNITED STATES CONGRESSMAN, BUFFALO, NEW YORK, 1971–89**

★ **AFL-NFL QUARTERBACK—PITTSBURGH STEELERS, NEW YORK GIANTS, SAN DIEGO CHARGERS, BUFFALO BILLS, 1957–1969**

Pro football gave me a good sense of perspective to enter politics. I'd already been booed, cheered, cut, sold, traded and hung in effigy.

—JACK KEMP

Ever since I was a kid, I thought about playing quarterback in the NFL, but I didn't tell anyone, because I was afraid they would laugh.

One day, my freshman football coach, Payton Jordan, who was a local legend in Southern California, called me into his office, which was awe-inspiring because it was a little like walking into a Knute Rockne shrine. He sat me down and said, "Look, out of all the players on my team, you're the one who I think can make it in the NFL. But you're not training enough, studying enough, lifting enough. You're just not putting in the work."

Suddenly I had someone besides me who believed I could be NFL material. I can't tell you how much that meant to me. It got me through the tough times and helped chase away any self-doubt.

In retrospect, I am not sure how many people he told that to, but at the time I thought I was the only one, and it meant the world to me. In reality, he was not worried about us going pro, he just wanted to make us better men. In order to do that, he knew he had to help shape a top-notch work ethic.

**HOW IT HELPED**

I was a physical education major in college and I was only concerned with football, doing just enough to get by to stay eligible to play. By the time my football career was over, I had grown to understand how important education was and what a passion I had for politics. After a while, I started to think of myself as a politician. I ran for Congress and won. Being a quarterback was the perfect training for any political leader because I knew how to work hard, how to inspire others, and was an optimist. Talk to any quarterback and they'll tell you they don't hope a play works, they know it will. They have to sell that play to their teammates, and I knew how to sell myself in politics the same way I sold myself on the field.

My greatest political achievement might have been helping to push President Reagan and President Bush, both forty-one and forty-three, into agreeing with my belief that lower tax rates would strengthen the economy and make it grow. I think my passion and research on the issue won them over, and a large part of that came from my years as a professional football player.

## FAMILY

I took great pride in helping to raise two football-playing sons who went pro. My son Jeff played for eleven years in the NFL, and Jimmy played for eight years in the Canadian Football League. With sixteen grandchildren, nine of whom are boys who all love sports, who knows, I might be back on the NFL sidelines again before too long.

## MY WRAP

*Jack Kemp always practiced full-contact politics, and it's hard to see him today and not think football player. We keep coming back to the same theme: one person, in this case Payton Jordan, taking an interest in a kid who needs a push to reach his full potential.*

# ELISABETH HASSELBECK

★ **COHOST OF** *THE VIEW*

★ *SURVIVOR: THE AUSTRALIAN OUTBACK* **FINAL FOUR**

★ **WIFE OF NEW YORK GIANTS BACKUP QUARTERBACK TIM HASSELBECK**

Be more concerned with your character than with your reputation, because your character is what you really are, while your reputation is merely what others think you are.

> —JOHN WOODEN,
> *legendary UCLA basketball coach*

I joined a softball league when I was nine, and my dad was so supportive that he got me this incredible glove. As soon as I had one practice, I knew I had a problem. I wanted to play shortstop, and they already had a shortstop. I was just obsessed with beating this girl out, and I finally found a way to do it, too. Not by manipulative means, but by earning it with the glove and the bat. I did not have to be convinced to work hard. I just kind of knew that's what it takes. And I was right.

I played softball through high school, but once I got accepted to Boston College, I thought that would be it. They had a top-flight Division I team and since I wasn't recruited, I didn't plan on playing. But somehow my dad convinced me to walk on. I came back from my first practice humiliated. I called him on the phone and just ripped into him, because I was in way over my head. I didn't get my bat on one ball and I knew I needed to show something of my game, so I made sure to display my speed. Thankfully, BC had and still has this incredible coach named Jen Finley, and for some reason she left me on the team. I remember when they were announcing the players who made it and she called my name my first reaction was to say, "Are you sure?" Now that I think of it, I did the same thing I did at nine that I did at nineteen.

To become the shortstop, I showed off my bat and how to make the cutoff. But my best asset was speed and Coach Finley liked speed, so I spent the entire season as a pinch runner. I learned to adjust to adverse circumstances, and I still use that quality today.

**LET THE SACRIFICE BEGIN**

In fact, I spent two years solely as a pinch runner. So there I was, getting up for 4:00 AM practices, going to all the games, working out in the off-season, playing an off-season schedule as a specialty player, and yet I still loved it. After two years, though, I was determined to crack the lineup, so I stayed at school in the summer, found a job and a hitting tee, and just hit and threw until my hitting got better and my arm got stronger. And it paid off, because I became the starting right fielder and was named captain of the team.

Senior year, I was expecting to make a real breakout, but I tore my rotator cuff and sat and watched the entire season. It was awful.

### WHAT I LEARNED

Because of playing college softball I found how important it is to never be late. If I was late for the morning practices, and I was, twice, the whole team paid the price by running the stairs as punishment. So we rarely ever slipped up, if only because we didn't want our teammates to suffer the consequences. Today, if I'm even a minute late for *The View,* I'm in a full sweat. Why? Because not only am I conditioned to be on time, I know how it hurts everyone on the team, or in this case, on the show, if I'm not there when I should be.

### THANKS, DAD

Overall, it was a tremendous experience and I owe it all to my dad. He was the one who got me out there and played with me. He was the one who got me the best equipment, even though we were far from wealthy. And I am just so thankful that he did. I know you're supposed to go to college and forget everything your parents taught you, but I never did. I always knew that he believed in me, and I thought I just couldn't let him down. He put so much value in my drive to play. He would wrap my glove and prep it for the new season. He instinctively knew what I grew to know, that I loved being part of the team. He knew that the discipline I derived from sports would help me in anything I did in the future. And he was right.

### SURVIVOR

There's no doubt that participating in sports helped me advance to the *Survivor* finals in Australia. I was trained to do whatever it took to be successful, and I also understood the concept of team. Both these qualities help you shine in *Survivor.* I was also able to outperform hundreds of contenders to join *The View* as their new cohost. I think it worked because I knew how to get along with my other hosts. I know that it's all about the team, not me, and in this case the show is one of the most successful ones on TV today, because of the team. I have to believe that I got the job and did so well on *Survivor* because of my years playing softball. I don't quit and will do whatever it takes to stay on the field.

**MY WRAP**

*If anyone understands the need to compete at any level at any time in any sport on any day, it's Elisabeth. She knows what it's like to be a pro, like her husband, Tim, who plays for the Giants, and her brother-in-law, Matt, who took the Seattle Seahawks to the Super Bowl. You just know that her daughter, Grace, has the kind of parents who will give her the right encouragement, drive, and outlook to make it in some kind of sport.*

# MICHAEL WALTRIP

★ DAYTONA 500 CHAMPION, 2001, 2003

★ OVER 120 TOP 10 FINISHES

★ NASCAR DRIVER, 1985–PRESENT

The problem with many athletes is they take themselves seriously and their sport lightly.

—MIKE NEWLIN,
*former NBA player*

## A RELIEF

When I was born, my brother Darrell was sixteen and was just beginning to make a name for himself in racing. As you know, he'd retire as one of racing's all-time greats. I watched his every move and naturally, I wanted to be like my big brother. I was not a great baseball player and basketball was not my thing, so when I decided on racing I just had to be a winner. I remember thinking, "What happens if I am not good at it?" When I finally got in the go-kart when I was twelve, I was good as anyone and I was relieved. In fact, I won my first big race. Turns out I had an instinct for it.

## SACRIFICE

I grew up in a small town. My mom worked at a grocery store and my dad worked at a factory. They worked hard and sacrificed so they could watch me race on weekends. I saw their passion and sacrifice up close and I just couldn't let them down, especially knowing they didn't have much money and were also giving me their time.

## WINLESS STREAK

By the time I finally made the Winston Cup Series (now known as Nextel Cup) I'd had some success, but I had gone four hundred sixty-two races without a win. There were times at the beginning when there was just no way I could win competing on a shoestring budget. Soon my equipment improved, I learned more and got more sponsors, and the opportunities started to come. Through it all, I still believed that given the right circumstances, if I could put myself with the right team at the right time, I could be a winner.

## FINALLY . . . BUT THEN . . .

Here I am with my first win at Daytona, the sport's biggest race, with my brother in the broadcast booth, and then I learn Dale Earnhardt had died. He was the guy who believed in me enough to put me on his team, and now he was gone. It's not like I had to put the sport in perspective because

I had it in perspective, but I would say that enjoying the win after that tragedy was almost out of the question. What I learned from Dale's death is a famous line, which I will paraphrase: "If I can't find what I am looking for in this world, maybe I was not built for this world."

## GOT TO BE GOOD TO LOSE

I had lost four hundred sixty-two races in a row, and I don't know if anyone can appreciate this, but you have to be good to lose that much and get in the car and race the next Sunday. Okay, so my numbers won't get me into the Hall of Fame, but that doesn't mean I'm any less of a person, or less of a dad to my daughter. Now I was starting to get beaten down by the people who questioned my lack of success. My patience actually got thinner with people who questioned me. I think people get caught up in things that don't matter as much to me. I've now won two Daytona 500s, and there are only five or six other drivers who've ever done that. I didn't believe I was as bad as they said when I was not winning races, and I don't believe I am as good as they say even after winning races.

## SLOW AND STEADY DOES SOMETIMES WIN THE RACE

I am thrilled I was able to survive and win the Daytona 500 after going through the drought. The losses built my character and made me more appreciative and respectful of others. I got angry and moody after losing, but I didn't let it affect my self-esteem. After losing, I had to get over it to do the things I had to do the next week in order to get into the win column. I'm proud that I did that every week and didn't change course or, worse yet, quit.

## NOT OVER YET

My best years are yet to come, and what I'm doing on the ownership side will show how innovative I can be. If you look at my numbers and say, "He wasn't very good," that's like looking at O.J. Simpson and saying he was a great football player, so he must have been a great person. But say what you want, you will not hurt my feelings.

## PEOPLE CAN RELATE

I think people appreciate the way I treat them. I get jazzed when I see people of all ages get excited when I appear. I see kids and older people just light up when I walk up, so I must be doing something right.

## IT'S NOT WORK

Even when you lose, this business is not work. As long as you're confident or competitive, you'll always find out why you didn't win. Your goal is to never quit, as long as you keep believing in your talent.

## MY WRAP

*Michael's a living, breathing example of someone who's a winner, even if he doesn't win as many races as his opponents. And he never gives up. He also has pride in knowing he was able to win the big races and never would have had he quit when things looked grim. Bottom line—people can relate more to the 0-for-462 then the 10-for-10. Just think, if he'd quit at 462, he never would have been 1-for-463. Michael's story is a great example of not letting the expectations and judgments of others decide your fate.*

## ARTHUR BLANK

★ OWNER, ATLANTA FALCONS

★ FOUNDER, HOME DEPOT

One day of practice is like one day of clean living. Doesn't do you any good.

—ABE LEMONS,
*longtime basketball coach, Oklahoma City University*
*and University of Texas*

I played baseball as a kid in Flushing, New York, and loved being the catcher, because I was in every play and I was directing the game. I still like directing things, being in control, which worked out well at Home Depot.

## THE MOMENT

I was seventeen and this kid was coming around third base and I caught the throw to the plate and he just barreled over me, flipping me backward. I popped up and threw the runner out trying to go from first to third. I'm sixty-three, and I still remember that play today. I can still see his face as he came around third. This guy wanted to kill me. I don't remember what the crowd did, but I do remember my team went nuts after I made the play. That was an amazing feeling. I loved being there for my teammates. It was probably the first time I got up out of the dirt, but as it happens, I would need that quality later in my career. I didn't always have Home Depot. I got fired from a home improvement retailer called "Handy Dan," and had to come up with something else to do. Maybe I wasn't in the dirt at the time, but my career certainly was.

## ANOTHER SPORT, ANOTHER MOMENT

One argument in which my dad took my side over my mom's was whether I could play football. My mom thought I was too small and my dad was all for letting me give it a try. He knew I was fast and perhaps tough, and he knew that's what you really need in that game. So I did play, and I'm so grateful that they gave me the green light. I played three years at Stuyvesant High School as a defensive back and wide receiver. Junior year, we had a great team. Thirteen guys got full rides to college. I couldn't crack the starting lineup until the biggest game of the year, when the starting defensive back got hurt. I knew I should play, but I wasn't sure if my coach thought I could do it because the matchup was pretty ominous. The kid I would be covering was the number-one receiver in the city and I was virtually untested. I will never forget what happened next. Joe Almonte, the captain of the team, went up to my coach and said, "Start Arthur Blank, he'll do fine," and that meant so much to me. It gave me the confidence I needed to step up, and I guess you could say I did, because this guy had fifteen passes thrown his way and caught only two the whole game.

## MY APPROACH TO PLAYING BALL AND BUSINESS

I want people who work for me to feel like they matter and feel free to make mistakes.

—ARTHUR BLANK

If you're going to play any sport at any level, you have to be willing to fail. If you don't fail, you'll never grow. If you're willing to fail, you'll always put yourself in situations where you'll excel. Teams that do well are willing to make mistakes and ultimately fail. This is my philosophy in business as it was in sports. As an example, I give you Tiger Woods of a couple of years ago. After all his early success, he took his game apart and rebuilt it. While he was doing it, almost everyone thought he was nuts. He took a risk by deconstructing his swing, but now he's better then ever.

## FINAL THOUGHTS

I love being part of a team that wins. I realized as a high school kid how much it means to have a coach, and now a boss, express faith in his employees. It lifts people up and makes them feel like they matter and, invariably, they perform better. It's a lesson I will never let go of. I'm always looking for someone who can do a job and do it well but who has not yet had the opportunity to show it. I love to give people a chance to step up. That comes from my days playing ball.

## MY WRAP

*Mr. Blank pushed to be a catcher and a defensive back to show that he is fearless. Somehow, he learned there is nothing wrong with failure. Amazing how much the sixtysomething business titan is like the kid growing up in Queens decades ago. It's probably why he gets along so well with his Falcon players as well as his employees at Home Depot. In Blank's world you'll never be criticized for trying something new, only for not trying at all.*

# MIKE MODANO

★ **U.S. OLYMPIC HOCKEY TEAM, SILVER MEDAL, 2002**

★ **7-TIME NHL ALL-STAR**

★ **STANLEY CUP CHAMPION, 1999**

★ **NHL CENTER, MINNESOTA NORTH STARS, DALLAS STARS, 1989–PRESENT**

Hockey players are like mules. They have no fear of punishment and no hope of reward.

—EMORY D. JONES,
*St. Louis Braves (CHL) general manager*

My dad, who worked every day of his life, taught me how to play sports. He was a construction worker in Detroit and I watched how hard he worked, how he went to work outside in all kinds of horrible weather with bosses few would like to work for. I knew I did not want to live that kind of life, and my dad did all he could to keep me from that life. He was not a vocal guy, but he modeled how I wanted to approach life. He would never let me be the one who pointed the finger or who told others what they should do. He taught me to be positive and to do what I could do.

Early on, I was in a lot of trouble. I was always getting into fights, not listening to the teacher. My parents put me into hockey to wear me out a little bit, to burn off some of that energy. I liked it so much, they then used it as a discipline tool. If I acted up or I got a bad report card, no hockey. It worked because I had such a love for the sport. I also had little energy to get into trouble at home or in school after an afternoon of practice.

I was doing well enough that at fifteen I got an offer to go to Canada and play, so, with my parents' blessings, I just packed up and left.

## MY MOMENT

I was worried that I'd end up back in Detroit doing construction with my dad. Until one tournament in Russia, I had no idea how I'd measure up internationally. But when I was named to the Junior Olympic team, I did well. After the tournament, the leading scouting bureau came out with their list of young prospects and I was up near the top. That's when I said, "Wow, this really might work out. I could actually have a career doing this."

## RAISED ON HOCKEY

When you leave home at fifteen you are thrust into many different situations and it's expected that you will grow up quickly. I did. I was so much older than my sister, it was as if I were an only child and I felt like I could get what I wanted when I wanted it. Joining a team, I quickly learned that wouldn't be the case.

The more I think about it, the more I realize how much about life I learned playing hockey. I was able to see and experience different countries and cultures. I saw how big business worked as we interacted with all our sponsors. I learned about all the different people and personalities from every corner of this country as well as from Canada, for that matter.

## MESSAGE TO THE KIDS

After doing that scene in the *Mighty Ducks* movie, kids seem more attached to me and that's okay. I get kids coming up to me all the time, telling me how they are going to be a pro player like me. I think it's great, but I let them know the amount of sacrifice that goes into it. I try to let them know about the hours away from the ice, how much training is involved in the off-season. They see the glamour of the glossy hockey cards and the highlights on TV, and I think that's cool. I don't want to discourage anyone, because it's a great life, but like anything else, it takes a lot of work. To make it, you also need a lot of luck. It's also clear to me that the process of refining your skills and expending the energy and sacrifice to be a pro is a wonderful teaching tool, even if you don't make it.

## HOW HE PLAYS THE GAME

I play the game with a quiet intensity. I try to keep my emotions in check. Yes, I have an intensity that matches anyone else on the ice, but I try to keep it low-key and under control. It's what works for me.

## WHAT'S NEXT?

I know I have just three to five years left in my hockey career, and I've often thought about what's next. I have sat in my accountant's office and gotten the sense of what his job is like. I think it will take the same type of drive to be successful in my post-hockey career as it did in my hockey career. I like to think I am ready for it, and I plan on bringing that quiet intensity to whatever I do next.

## MY WRAP

*Mike has sacrificed his entire life to be the best hockey player he can be, and in return he's had some incredible success: All-Star games, Stanley Cup wins, and many, many more thrills. As he admits, his life was heading south at a young age, but getting involved with sports truly straightened him out. It should be interesting to see if the star with the matinee idol looks will, indeed, win another Cup, or maybe start another life in management in the sport that he loves.*

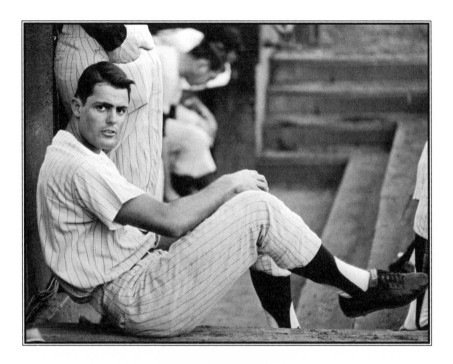

# LOU PINIELLA

★ **MANAGER OF THE YEAR, 1995, 2001**

★ **WORLD SERIES TITLE, CINCINNATI REDS, 1990**

★ **BASEBALL MANAGER, YANKEES, REDS, MARINERS, DEVIL RAYS, CUBS**

★ **2-TIME WORLD SERIES CHAMPION, 1977–78**

★ **AL ROOKIE OF THE YEAR, 1969**

★ **MLB OUTFIELDER, ORIOLES, INDIANS, ROYALS, YANKEES, 1964, 1968–1984**

Now, there's three things you can do in a baseball game. You can win, or you can lose, or it can rain.

—CASEY STENGEL,
*Hall of Fame baseball manager*

I was having some success moving my way up in the lower minor league level, but that changed when I hit Double-A, in Elmira, New York, where I didn't hit higher than .250. I was a pull hitter and I just never hit the ball to right field. So I was traded to the Cleveland Indians and I was sent to Portland Triple-A, where my season took on the same pattern: good power numbers, but low batting average.

## MY MOMENT

In 1966, my manager pulled me into his office and said, "Either you use the whole field at the plate or I will send you back down to Double-A." And then he added something that was so very important. He said, "I'm willing, Lou, to come out and work with you every day before the game. And during the game, you will not be permitted to pull the ball. If you pull the ball, even if it's a hit, I'll sit you out the next game." So he got me a bigger bat and we would work every day until I learned, for the first time, how to hit the ball to right and center field, using the whole field. I also learned to recognize a slider and a curveball. Looking back, if it were not for this one manager taking an interest in me and making me a more complete player, I would have floundered and I would never have become a major leaguer.

## LOU GROWS UP

The difference between making it and not making for so many players is often what they do and don't do off the field. Going pro after high school left a lot of room for me to grow and I knew it. I tried to pair up with the most mature player on the team, the guy with his head screwed on straight. See, you have a choice to run with the wrong crowd or run with the right crowd, and I chose the right crowd. The player I chose knew when to have a beer and when to hit the cage, so I mirrored his schedule and it really worked out for me. So many more talented players failed because they could not say *no* to the nightlife.

## MY MESSAGE AS A MANAGER: FOCUS

My message to my players has always been to focus on baseball. Forget the restaurants, the clubs, the car dealerships. There will be a time in their lives for that, and while they're playing is not the time. I also push my players to finish their degrees in the off-season. You also have to like people. Not everyone is going to be like you, but you have to learn how to get along with all types of people. I became a people person because I wanted to be successful. To do that you have to see how your players tick. More than anything else, I enjoyed watching players grow before my eyes. When it gets to the point where they get the big contract or win a championship, well, that's the definition of rewarding to me.

## OFF THE FIELD AND INTO THE BOOTH

When I'm in the broadcasting booth, I pour myself into this job like I did as a player and a manager. I won't compare myself to Joe Buck or Tim Mc-Carver, who are the best right now, but I will look to be myself and put in the time so I have something sensible to say. Keep in mind, I also think it's important to be entertaining, so I will always try to inject some fun into what I have to say.

## WHAT IF BASEBALL DIDN'T PAN OUT?

I think I would have figured it out and been okay. I consider myself a motivated guy, and I love challenges. If my career was tied to financial success, I think I would have done well. But I'm just glad I didn't have to!

## THE WRAP

*Lou has done what most thought impossible: he's had two incredibly successful careers, first as a player, then as a manager. Like so many others, Sweet Lou needed someone to care about him, or else we never would have seen him on the major league stage. Maybe you're in a position to help someone as well, and who knows, possibly even change the destiny of a young life.*

# BILL GOLDBERG (AKA "GOLDBERG")

★ ACTOR, *THE JESSE VENTURA STORY, THE LONGEST YARD, READY TO RUMBLE*

★ WWE AND WCW WORLD CHAMPION WRESTLER

★ DEFENSIVE LINEMAN, ATLANTA FALCONS, 1992–1994

Ability may get you to the top, but it takes character to keep you there.

—JOHN WOODEN,
*legendary UCLA basketball coach*

I jumped into football in eighth grade. I grew up in Tulsa, Oklahoma, and I didn't start playing before that because there was a concern that I'd break something or somehow affect the growth process. Even though I did play some baseball and basketball, I never liked it as much as I did football. But even if I had, when you have two older brothers playing football, I was pretty much programmed to follow in their footsteps. Not only did I think it was possible for me to go pro, I actually thought it was probably because my brother was a college star. Not to sound too cocky, but I always knew that's what I'd do.

## SINGLE-MINDED FOCUS

The breakout year for me came when I was a junior in high school. I started to get some interest from colleges, and when I was a senior I was recruited as a lineman by the University of Georgia. I became an All-American nose tackle and was good enough to get drafted by the NFL.

My goals were coming true because I was obsessed with being a football player. You've heard of Todd Marinovich, the former number-one pick of the Raiders whose dad ruined him by forcing him to play football every day of his life. Well, I was my own Marinovich. I was like a cyborg. I didn't want to drive my car across the finish line. I wanted to be the hood ornament. I was not always the best guy on the field, but I was the guy with potential who worked really hard to attain his goals.

## HOW DID HE LEARN THAT HARD WORK PAID OFF?

We had good, basic values. We knew what it meant to work hard, and I never expected to get anything for free. I credit my parents, because they always made me work for anything I got. When you do that, you get a degree of pride that's hard to define. Still, having said that, I did not reach my goals in football even though I may have maxed out my ability. I wanted to have the impact of Jerry Rice, Joe Montana, and Dick Butkus, but it didn't happen. In reality, I was the eleventh-round draft pick of the Rams, played five years, and then an injury shelved my game forever.

*I think being one of twelve hundred guys in the nation to play in the NFL is a huge success, but obviously Bill doesn't think that way.*

The problem with having just one goal is that you don't have a second goal in the event that you don't reach the first one. When I got hurt, perhaps it was the best thing that happened to me because it forced me to move on. I was under the belief that the next year, or the year after that, would be the year I broke out and went All Pro. But, of course, there was no next year. I retired and then just chilled out for about a year until it was time for Plan B—wrestling.

*The downside to having an unwavering belief is that more often than not there is no fallback plan. It took a year, but Bill got one, and in this case Plan B turned out to be better than Plan A.*

Maybe it was in the cards for me to be a wrestler. My college roommate, Kevin Greene (a linebacker with the Rams and Panthers), used to tell me that I would make a great professional wrestler. My brother roomed with a professional wrestler. Professional wrestlers owned the gym I worked out at, and they just pushed me to do it. I got a great start wrestling for the now-defunct WCW, which was a lot of fun. It was a whirlwind experience and I was just in the right place at the right time with right look, and the whole thing just took off. When the money and the fame came, it was cool. Not that I wanted the acclaim. I just wanted the power to impact people's lives. I also wanted to make my parents proud. But with the wrestling came the notoriety with the kids, and I loved having them look up to me. Before every match, I made it a point to visit the Make-A-Wish Foundation in every city, and those kids looked up to me like I still look up to Joe Montana. Today, I spend a lot of time meeting the troops and they seem to love it. For me, it's a thrill, a gift that I cannot begin to describe. But I hope it never ends. And even though I'm now doing some acting and have a degree of fame from that, make no mistake about it, I owe wrestling for my high profile.

## ON BEING JEWISH

My faith does not dominate my life, but every day, when I wake up, I am proud as hell to be Goldberg, proud as hell to be Jewish.

## FOOTBALL PLAYER FIRST

I may have had much more success as a wrestler, but in my mind I am a football player and I always will be. If I have to introduce myself to someone who does not know me, I always introduce myself as a football player and not a wrestler. Until I'm one hundred ten years old, I will still be a football player.

## MY WRAP

*For a guy to be this focused and this talented and still not reach his predetermined goals should only make us feel better if we should fall short. Bill's work ethic in sports gave him a template to excel after sports, and he was determined enough to make the next dream work, even if the first one didn't.*

# BETH OSTROSKY

★ **SUPERMODEL**

★ **TV AND MOVIE ACTOR**

I would like to deny the statement that I think basketball is a matter of life and death. I feel it's much more important than that.

—LEE ROSE,

*University of North Carolina at Charlotte basketball coach*

I was captain of the Fox Chapel High basketball team in high school. I had many college scholarship offers, but instead I decided to capitalize on my modeling career, which was just beginning to take off at the time. I'm happy with my decision because I am still modeling today and I know I would not still be playing basketball.

## THE "BLOODY" MOMENT

We were not the best team, but we worked hard. As the power forward, I felt like a lot of the scoring was on my shoulders. I will never forget going up against the first-place team. I was matched up with this 6'2" German exchange student. She was probably the best-known player in the state. Well, I went up for a jumper and she elbowed me in the face and my two front teeth fell out of my head. Blood was pouring out of my mouth and I was in shock. My dad was at the game and he happened to be an oral surgeon. He came out of the stands, picked up my teeth, pulled my head back, stuck them back in, and squeezed the bone together in the roof of my mouth. I looked at him, he looked at me, I nodded, and I went back into the game. We lost by twenty points, but I never felt so good.

*Keep in mind that she was just beginning a modeling career, and there's not much call for toothless cover girls.*

It hurt like hell and I swallowed a gallon of blood, but I kept my tongue against my teeth to set them in place. What's great about this story is that my teeth roots caught and they are actually as they were originally, even today. Of course, through it all, I still remember my mom crying. I thought it was for my welfare, but it was actually because she saw my modeling career pass before her eyes.

## WHAT I GOT FROM HOOPS

I model like an athlete, meaning I treat the modeling business as I did team sports. I show up on time. I show up sick. I stay late. It's just in my blood from my days on the court. Today, I train hard and I never worry about a nail breaking or a finger fracturing. I just go after it. Basketball and sports are in my blood. As soon as I could walk, my dad had a ball in my hands. Knowing my family was in the stands meant everything to me. I feel so sad for people who were not in organized sports because your whole sense of self comes from sports in almost everyone I know. Howard (Stern, her boyfriend) never played organized sports, and it kills me because he would have been a great teammate. If you watch his show and see how he interacts with his staff, you'll see that he's just a natural captain, even if he was not a natural athlete.

## HOW BETH PLAYS THE GAME

I played with my heart and soul and I always played and practiced hard, even practicing on my own. My days in sports, especially that game where I lost my teeth, gave my dad that twinkle in his eye, and I would not trade them for anything in my life.

## MY WRAP

*This story is almost incomparable. The career she chose, despite her obvious good looks and poise, is absolutely brutal. It's hard to imagine her days in sports not helping her break through as she has. By the way, when she gets on Leno or Letterman, she plans on telling the same story she just shared. So next time you see her on camera or in person, stare at her teeth.*

# THEODORE ROOSEVELT

- ★ 26TH PRESIDENT OF THE UNITED STATES, 1901–1909
- ★ NOBEL PEACE PRIZE, 1906
- ★ U.S. VICE PRESIDENT, 1901
- ★ GOVERNOR OF NEW YORK, 1899–1901
- ★ COMMANDED FIRST U.S. VOLUNTEER CAVALRY REGIMENT (ROUGH RIDERS), SPANISH-AMERICAN WAR, 1898
- ★ ASSISTANT SECRETARY OF THE NAVY, 1897–98
- ★ NEW YORK CITY POLICE COMMISSIONER, 1895–1897

In life as in football, the principle to follow is hit the line hard and don't foul and don't shirk, but hit the line hard.

—TEDDY ROOSEVELT

SOURCES: TWEED ROOSEVELT, GREAT-GRANDSON; JIM FOOTE, ROOSEVELT HISTORIAN

This president ended up on Mount Rushmore, so I think you can conclude that historians think Teddy Roosevelt was not just another President or an average leader. As robustly as he lived his adult life, he began it as a child riddled with asthma and often had to sleep sitting up in a chair in an effort to breathe. He was virtually a house prisoner. Most people did not expect him to live through his teens—that was how bad the asthma was. But he survived to live an adult life that was the complete antithesis of his youth. When it wasn't full-contact politics he practiced, it was full-contact sports.

> *"My exercise consists of walking, boxing, vaulting, and gymnasium work."*

Eventually, young Theodore's father built a gymnasium for him where they lived in Manhattan and said, "Son, you have the mind, not the body. One has to be built up, and it's hard drudgery but I know you can do it." He was so frail as a kid that he had to be home-schooled. He spent countless hours on the parallel bars and with dumbbells trying to build up his chest. Physical exercise, it turns out, was and is one of the best treatments for asthmatics. While he was bright and had a photographic memory, he was not able to socialize. He later admitted that he was afraid of a great many things, but found that by practicing being brave, he became brave.

> *"I believe that those boys who take part in rough, hard play outside of school will not find any need for horse play in school."*

**TEDDY TURNS**

Part of the treatment for asthma back then was getting out of New York City. Come summertime, Teddy was shipped out. He took the train up to Moosehead Lake in Maine.

On the stagecoach ride from the railroad depot to the lake where he would be staying, he encountered a problem. Two much larger boys around his age started in with him. With his quick temper, he swung out at them. The result was that the undersized Teddy was manhandled. Without any effort at all, the other two boys held him off while he flailed away like a windmill, never landing a blow. He was humiliated and vowed never to let it happen again. He got his dad to put him in boxing class.

EXCERPT FROM *TEDDY ROOSEVELT: YOUNG RIDER*

## HIGH-VALUE FIGHTER

At fifteen, his dad was able to get Teddy a boxing coach named John Long. He was still just one hundred thirty pounds, with short arms, and nearsighted, but his asthma was virtually gone. Like everything else he did, Teddy picked up the skills quickly and learned to hit straight, hard, and clean.

## TOURNAMENT TIME

It was soon time to test his skills in the ring as a lightweight in Coach Long's boxing tournament. Roosevelt drew a tall, slender fighter in Dick Williams. It's been said that he never hit Williams hard but did hit him often and easily won the fight. He went on to take his next two bouts. The fourth was for the trophy. His opponent was John Hart. Like almost everybody else, he was taller, with longer arms, than the future President. After two rounds, with the fight all even, the bell sounded. Teddy dropped his hands. Hart, already in motion, whacked him with a full shot right in the face. As much as it hurt, he knew that Hart did not hear the bell or at the very least did not mean to hit him after the bell. So, before the ref could stop it or the crowd would go over the top, he yelled, "Quiet, John didn't hear the bell." That stopped the booing and kept the fight going. It was Teddy who held on to win the decision and the trophy. He was quoted as saying between rounds right after the blow, "You play fair and if the other fighter means to play fair, you stand up for him."

### IMPACTING THE GAME

In 1905, eight people were either killed or wounded on the football field. As President, Roosevelt felt he had to take action, but at the same time he knew the value of having athletics complement academics. He called the three big football powers at the time to Washington, D.C.—Harvard, Princeton and Yale—and they hashed out the National Football Rules. That coalition stuck together and expanded to become what we now call the National Collegiate Athletic Association (NCAA.). The Theodore Roosevelt Award is awarded annually for national significance and achievement in competitive athletics and attention to physical well-being in college. This award sums up what this book is about.

### FUN AND FISTS IN THE WHITE HOUSE

In 1906, Roosevelt was sparring with a navy lieutenant in the White House, got hit in the left eye, and suffered a detached retina. Back then, anything like that happening to a President would send the country into a major panic, and so it was kept quiet. Roosevelt lost his sight in his left eye and his doctor advised him to give up boxing. Roosevelt's reply was, "Very well, so I'll try jujitsu." He did, and eventually he earned a black belt.

### MY WRAP

*Although sports were not the single key to one of America's most accomplished citizens, Teddy Roosevelt went out of his way to say they were vital to his development. But more than winning, he talked about fairness and fitness. One of the hardest parts of writing this book was keeping this particular chapter brief because this man did more in his life than just about anyone else in history.*

# ANTONIO FREEMAN

★ CAUGHT 477 PASSES FOR 7,251 YARDS; GAINED 1,007 YARDS
  RETURNING KICKOFFS AND PUNTS; SCORED 64 TOUCHDOWNS

★ PRO BOWL, 1998

★ SUPER BOWL CHAMPION, 1997

★ NFL WIDE RECEIVER, GREEN BAY PACKERS, PHILADELPHIA EAGLES,
  1995–2003

It's amazing what the human body can do when chased by a
bigger human body.

—JACK THOMPSON,
*former NFL quarterback*

### THE DIFFERENCE ONE PERSON CAN MAKE

Anytime you make the jump from high school to college and then from college to the NFL, you have doubts. Packers head coach Mike Holmgren was the difference-maker for me. He didn't look to me as a third-round pick, ninetieth overall, and think, "He better prove himself to me." Instead, he pulled me aside early in my first camp and let me know I could be a great one in the NFL. Here I am wondering if I can make it in this league and this championship coach tells me I can be special! This is the guy who shaped Jerry Rice, John Taylor, and Sterling Sharpe and he was saying I could be all that. Those words from that man made my career. He didn't look for what I didn't have. He wanted to utilize what I did have. It wasn't long before I was working my way into the starting lineup as a wideout and returning kicks right off the bat.

### MOTIVATED EARLY

My parents expected a lot of me in school in terms of academics as well as in sports, and they would use incentives to make sure I focused because I was like any kid who had to battle distractions. I get good grades, I get the Jordan sneakers, the bikes, and a lot of material things. I don't get good grades, I don't get anything.

### HOW NOT TO DO IT

My family never told me that I was going to be a pro player and that it was the only way to fame and fortune. Even though I was recognized as a hoops standout and played well on the football team, I knew there were no guarantees. Why? Because my brother was one of those can't-miss kids who looked like he was heading to the pros. He was an All-State running back in Baltimore and he made some choices that messed up his chance to play major college ball, thereby destroying any hope of playing in the NFL.

He's six years older than me and my best friend in the world, but I just saw how he blew it. My parents and I learned from my brother.

They'd say to me, "If your grades are not high you're not playing and I don't care if your coach calls the house day and night. No sports if school suffers."

## LEARNING THE MEANING OF WORK: THE TEST

I went to Virginia Tech on raw ability. I was coming out of high school as the Maryland offensive player of the year and I was flying high, ready to start as a freshman. I never did speed training, nor did I lift weights. I was all natural, no work, which is the rap on Baltimore athletes like Sam Cassell and Carmelo Anthony. In reality, I was not ready and didn't play much. What was worse, I did not study and got all D's. My coach read me the riot act and then my dad got to me and let me know I had one more semester to get it together or it was over. Well, I did get a 2.5, got my body ready to play, and I turned it around for my second year, saving my career and my education.

## TEST PASSED

I ended up having a fine college football career and better yet, I got my business degree as well. I know now that even if the Packers hadn't given me the shot I would have been okay. I had that degree and I had my parents to ensure my success and happiness. Even today, as a so-called millionaire athlete, I still go back to my parents for advice and, of course, they sometimes still offer unsolicited advice.

## THE GREATEST CATCH IN *MONDAY NIGHT FOOTBALL* HISTORY

> "It's all about not giving up."
>
> —ANTONIO FREEMAN

I was having a mediocre game, about four catches, when my favorite play—a slant—was called. When I got to the line, I saw the Vikings were about to blitz, which meant I would look at quarterback Brett Favre and change my pattern from a slant to a slant-and-go. Well, Favre tossed it up, and I collided with the Minnesota defensive back. He thought the ball was on the ground, but all the while I felt the ball on my back. I turned around, grabbed it, and ran it into the end zone for a score. Favre jumped me and asked if I caught it. I said, "Yes," and he just couldn't believe it.

## UNSELFISH

I trained my replacement in Green Bay—Donald Driver. I wanted him to know what I was doing, so if I went down he could pick up the slack. I saw that he had great speed and I was able to teach him what I knew about the system, the game, and Favre. I got that mentality from Brett, because that's the way he treated me when I broke in. He made me feel like I mattered, and I was inspired by that. Can you imagine this Hall of Fame quarterback breaking down film with this third-round rookie pick? That's Brett. I wanted to pass down that wisdom even if it wasn't in my career best interest for me; it is all about the team's interest. Robert Brooks trained me to take his spot and I did take his spot, the same way Sterling Sharpe mentored Brooks, who would take his spot. In the end, it worked for everyone because we didn't quit doing the things that made us pros.

## WHAT I MISS

Since retiring, I miss the structure in my life and of course I miss the guys. I do not miss the hitting or the games as much. What I learned in my years in the game is a huge help to me today. The feeling of teamwork, deflecting credit, pulling for someone else, and preparing for any meeting like I did for a game certainly helps me out today.

## MY WRAP

*Parents matter more than anything else, and Antonio had great ones. He stumbled, but he reacted and adjusted to become an outstanding pro. For a time he was among the best in the league. Interesting, isn't it, how bad examples can be as great a learning tool as good ones, if taken right. That's what happened with his brother blowing his Division One football career. And it's also interesting that Freeman thanks his first NFL coach for his stellar career.*

# SEAN ELLIOT

- ★ **FIRST ORGAN TRANSPLANT PATIENT TO PLAY PROFESSIONAL BASKETBALL (MARCH 14, 2000)**
- ★ **NBA CHAMPION, SAN ANTONIO SPURS, 1999**
- ★ **2-TIME NBA ALL-STAR**
- ★ **NBA PLAYER, 1989–2001**
- ★ **NCAA PLAYER OF THE YEAR, 1989**
- ★ **ALL-AMERICAN SELECTION AS A JUNIOR AND SENIOR AT THE UNIVERSITY OF ARIZONA**

Playgrounds are the best place to learn the game, because if you lose, you sit down.

—GARY WILLIAMS,
*basketball coach*

I was never a dirty player. I always tried to stay within the rules, and I kept my mouth shut on the court. I could trash-talk with the best of them, but I never did it unless I was called out.

## PLAYING WITH MEN

We had two courts at the YMCA in downtown Tucson, one for adults and one for kids. The adult court was filled with former pros and college players and they would go hard. I had to prove I could play with those guys, or else I'd have to stay on the kids' side. That's how I learned to compete hard and be tough on the court.

## WHATEVER IT TAKES

In high school, I knew I had to shoot and score a lot for us to win, and I think I averaged thirty points a game. In college, it became a different story because we had plenty of guys who could score. Early on, I was not even the best player on my team.

## SEAN DOUBTS SEAN

I remember watching an ESPN special on the Five-Star Basketball camps with my mom and we decided I just had to go there. When we began to ask around to find out how to get in, we were told, "Why bother? You're from Tucson. You're not going to play big-time basketball, and you won't go to a big-time college power." I ended up going and thriving at camp, as I got noticed for my play.

The great moment for me was hearing Rick Pitino talk to us about working hard, maintaining the work ethic, and how to separate ourselves from the pack. I took that to heart and worked nonstop on my game. My mom was so tired of driving me to the gym that she just bought me a bus pass to go on my own. My friends thought I was crazy, but I knew it would take extra effort to become a player and I was willing to give that effort.

## MORE THAN HOOPS

My freshman year in high school, there was a feature story on me in the newspaper about how I played all sports. I ran track and played soccer, baseball, and, of course, basketball. It seemed like I could excel at all sports. But if there was any sport where I was having little success it was basketball. As a freshman and sophomore, I played junior varsity, but I hardly stood out.

## THE MOMENT

We lost our junior varsity opener by fifty-three points. But it got worse. I was working on a spin move, got twisted up, and shot in the wrong basket. It was *sooo* embarrassing, but it helped me grow as a person. After my freshman season on the worst JV team in the city, I turned to soccer. First game, I got slide-tackled and blew my knee out. I was in a cast from my hip to toe for two months and doctors told me I would never play again. After I got the cast off, I was told to stay off it for a couple more months, but I was in the gym the following weekend. The thought of the rest of the city getting an advantage on me while I healed was galling. What it did was make me redouble my training, concentrating on improving my shooting and building my leg strength.

## POWERFUL WORDS

When my sophomore season was set to begin, I was more than ready to play. I remember after one game in particular, my teacher's boyfriend came up to me and said, "Sean Elliot, you will be a player, a real star." It made me feel fantastic. Considering how far I and our team had come, well, I was just sky-high.

## TRANSPLANT

After the 1993 season, I was feeling sick. Blood tests showed something was wrong, but the doctors didn't know what. I took medication for six years, feeling sick on and off until 1999. I ended up needing a kidney transplant. My brother stepped up and gave me his kidney. My brother had helped me so much in so many different ways, this was just the most profound example.

I felt so good so soon, I would have trained in the hospital bed if I could have. I think I was the first person in sports to play with a transplanted organ, and it was wonderful to be able to inspire people who were going through a similar ordeal. By playing, I let them know that they, too, could make a full recovery. The doctor told me I couldn't make a comeback, the same as I'd been told back in high school, but I wanted to prove him wrong, too. I did.

## FINAL THOUGHTS

I owe my attitude and cohesiveness to my mom. She did everything for me and my brother. I watched her work the graveyard shift at the hospital after her divorce. She'd leave at ten-thirty at night and we'd be on our own. My dad would help, but it was all on her shoulders. I didn't want to let her down and learned to be really close with my brother. It was then and there that I learned I would do whatever it took to be successful. If my mom could sacrifice for me, I would certainly sacrifice a personal life for my career.

## MY WRAP

*Sean always appreciated the opportunities he earned, and he was the first to tell me he was almost too coachable. We don't often hear about being too coachable, but he gave up his athletic assets for Coach Larry Brown with the Spurs and Lute Olsen in college to conform to what the team needed. He later learned to keep his game and adjust it rather than change it altogether. He has put the same time and effort into broadcasting. If you like his work it's because he spends the night before watching game tapes to make sure his comments help explain the game you're watching.*

# ERIC BRAEDEN

★ ACTOR BEST KNOWN FOR HIS ROLE AS VICTOR NEWMAN IN THE SOAP OPERA *THE YOUNG AND THE RESTLESS*, FOR WHICH HE WON A DAYTIME EMMY IN 1998

★ APPEARED IN *THE RAT PATROL, THE MARY TYLER MOORE SHOW, ESCAPE FROM THE PLANET OF THE APES* AND *TITANIC*

The rules are very simple. Basically it's this: If it moves, kick it; if it doesn't move, kick it until it does.

—PHIL WOOSNAM,

*North American Soccer League commissioner*

As you know, after World War II, Germany was a mess. But as kids, we still tried to make the best of it and for us that meant playing soccer. If we didn't have a ball, we would find our way to the butcher and ask for a pig's bladder. We'd surround it with straw and then one of the moms would sew cloth around it and we had a ball and the game was on.

From the end of the war in 1945 to 1948, we were all dirt-poor and yet somehow soccer survived. We often played in our good shoes, or should I say, our only shoes. Of course, when we got home our parents would have our heads. If I see a ball anytime, anywhere now, in an airport, in a parking lot, I'll ask for it so I can touch it, kick it, or throw it.

As much as I loved playing, the biggest moment in sports happened for me in 1954 as a fan and a German citizen. It was the first World Cup after the war and West Germany beat Hungary. I still remember crying after that game. It gave birth to the West German economic miracle. After that, no one had to tell me about the power of sports. I was a witness to it.

As a hockey player, I ran into a roadblock as a kid because I had about eight concussions—few wore helmets back then. The doctors would not let me play soccer, hockey, or anything else. This led me to track and field and I won a national youth championship the year before I left for America, which was one of my most memorable sports moments. The second most memorable moment happened when our soccer team won the national championship in 1972–73, when I was thirty-two. As much as I've done in show business—appearing in movies and on Broadway—winning that title was one of the happiest moments in my life. As we made our way through the season, if I had to miss a game, I would make a call back home to see how we were doing.

As for me, I played right fullback and also took all the penalty kicks for the team. In five years I never missed one.

*The man obviously loves pressure.*

I've always been like that in all facets of my life, and there's an excellent chance I got that from playing soccer. When the pressure is on, *I am on!* But I digress. Back to the game, which took place in front of a packed house. My focus was tremendous and I didn't sleep the night before. I scored the first goal on a penalty kick and we went on to win 3–1, and we were national champs. It was the best moment ever. All I care about is that we won as a team and you have to feel sorry for my poor wife, because this is what I talk about all the time.

## BRAEDEN THE PLAYER MORPHS INTO BRAEDEN THE COACH

I went right from playing to coaching soccer. My son, with whom I'm very close, joined a soccer team when he was five years old and I coached him and his friends until they were thirty. Twice we made it to the final four in the national championships. I find coaching to be one of the most exhausting things you can do, and I have enormous respect for anyone who does it.

## WHAT I GOT FROM IT

One thing I learned from both playing and coaching is to never give up. Sure, my wife and I hated Sundays after a loss, but by Tuesday I was ready for the next game, and by Thursday I knew we were ready to win on Sunday.

I learned that sports, like life, offers second chances, if you're willing to take them. You can always try harder to be better the next time around. My experience with sports taught me to have enormous tenacity, and in TV and the movies you really need it. It's true that my parents and brother probably helped build this drive in me, but sports certainly honed the drive and discipline and provided me with a way to apply it. I learned how to get better in anything I would approach. I still remember finding a wall and hitting it for an hour with just my left foot, simply because it was the only way I knew to improve my soccer skills.

I've been on forty-five covers of *Soap Opera Digest* and in 2004 I got my latest Daytime Emmy nomination for best leading actor and I still feel like I am at the top of the heap today. Why? Because I work at it and I don't take any crap. I demand respect and I give it. This is something I got from playing sports.

**HOW ERIC PLAYS THE GAME**

I would say in one word, fairly. If I get beaten I will congratulate you and if you are too cocky I will come after you.

**MY WRAP**

*Eric is as passionate about the power of sports as anyone I've ever met, and I interviewed around two hundred people on the subject of sports. He's prideful yet modest. In my mind, from sports to acting to business to politics, there is nothing this man would not have excelled in.*

# GALE SAYERS

★ MEMBER OF THE 75TH ANNIVERSARY ALL-TIME TEAM

★ 3-TIME NFL ALL-STAR GAME MVP

★ ALL-TIME NFL KICKOFF RETURN LEADER

★ NFL RECORD 6 TOUCHDOWNS IN ONE GAME (VS. SAN FRANCISCO), 1965

★ NFL ROOKIE OF THE YEAR, 1965

★ NFL HALFBACK, CHICAGO BEARS, 1963–1971

You don't develop good teeth by eating mush.

—EARL BLAIK,
*Army football coach, 1941–1958*

I felt I could do things a little faster and a little quicker even at the age of eight, playing midget football. I guess I felt I could see things a little differently than the other kids I was playing with and against.

## THANKS, COACH

My major success comes from what I learned from coaches as a kid. For example, when you go around the left end, hold the ball in your left hand so you can use the stiff arm. Even today, runners don't know how to do that, even in pro ball.

People ask me today how I get my meetings to start on time and why I am so rarely late and I say, "Thank my football coaches." If I was as much as ten seconds late for a practice, not only wouldn't I be allowed to practice, but I wouldn't be allowed to play in the next game. But I was also rarely late because my dad would never let me be late or allow me to talk back to my elders. It was all about respect. My coach backed up my dad and my neighbor backed up my coach and my dad. We had such a sense of community back then, it was almost impossible for my life or career to go south.

## HOW DID YOU HANDLE THE CHILD STAR STATUS?

I was never given star status as a kid because part of understanding the game is knowing that you don't score without blockers or without a quarterback, and you don't win without a great defense. So I was not looking for nor did I receive special treatment. I never thought about going pro. I never even thought about playing high school football until I got there. I never thought of playing college football until I actually joined the team. And I certainly never thought about the NFL.

## HE ALWAYS BELIEVED

Ten kids from my hometown died in Vietnam. I wasn't thinking about the pros; I thought I was going to war until my dad and coach told me to try college, where they said I could play at the next level. My dad was a real straight talker, so when he said I should try college ball, even with my mediocre grades, I instantly agreed to try. As you know, I went to Kansas (*where he became the Comet*) and then got drafted. By the Bears, not by Uncle Sam. If it wasn't for my dad and my high school coach, I would probably be dead in some Vietnam field.

## HOW I KNEW I BELONGED

Even when I got into Kansas, I still wasn't sure I could be a college player until one freshman varsity scrimmage. Our defense was more than a little mean and no one back on our team wanted to run the ball against them. But I just knew that I could be effective, and seeing the others shrink from running against them made me, for some reason I'm not even sure of, more confident. It was a similar thing with the Bears. They drafted me number-one, but I wasn't sure I could be a pro runner until I went to practice. I saw other running backs who were good, but I thought I was quicker, faster, with better field vision. I knew that if they gave me a chance, I could do this. In our first game against the L.A. Rams, I ran back a kickoff for ninety-seven yards and a punt return for sixty-five yards. I also threw a pass for a touchdown. Then I knew for sure that I belonged.

## STEPPING IN FOR DAD: PAUL HORNUNG

Paul watched me play in a game in which we lost 23–14 and came up to me and said something I will never forget. He said, "Gale, if you continue to work hard you could be one of the best backs to ever play this game." To hear something like this from a guy like Paul, well, I will never forget it and neither will he.

## PREPARED FOR THE UNEXPECTED

I knew how we looked when we practiced. I knew what my blockers had to do. My questions were more along the lines of, what happens if they don't make that block? That's why I learned to always be ready for the unexpected. When things went wrong, I had to make it work, not let myself get tackled and take a loss. I would call on my peripheral vision and try to see ahead of the play.

### LEARNING TO COME BACK/LEARNING TO QUIT

My first real injury came in 1968. I came back from it, though, and led the league in rushing. The second injury came in 1970, and that one was more serious. After that, I just never felt right. I told my coach that I would try to run in one exhibition game and if I didn't feel right, I was through. I ran the ball three times and fumbled once. At halftime, I said, "It's over."

From the day I played, I was prepared to quit. I made myself ready. I studied to be a stockbroker in the off-season because I knew I had to have a job because then we weren't making the money they make now. The highest salary I ever got playing football was sixty thousand dollars.

### SAYING GOOD-BYE TO BRIAN

As I mentioned, I lost a bunch of friends from high school in Vietnam, so the concept of death was not new to me. But Brian Piccolo's sickness and death at twenty-six took me and the nation by surprise. Four months after that, my mom died, and so it might have been the toughest year of my life. Walking away from the game was tough, but compared to watching your best friend and your mom die, it doesn't even rate.

### FOOTBALL OVER. GAME ON!

I had success in business after football because I worked just as hard at business as I did football. I wasn't looking for any easy job and there wasn't one waiting for me. I got my stockbroker's license when I was playing and I made it work. I had more talent in football than stocks, but like I did when I was in pads, I hit just as hard on my research and analysis. I know I probably had more natural talent in sports, but so did thousands of other people—people who know me know how hard I worked to be successful and how hungry I was. Gale Sayers was never looking for any star status. Ever. And after Wall Street, I was able to become the first black athletic director in the country.

### GOALS NOW

No one has given me anything because my name is Gale Sayers. I do well with my computer company because I have a better product and give better service. I'm not trying to be the next Bill Gates. I just want to work hard and see how far I can go. It's all the same—sports and business—it's a team game: the sales team, the marketing guys, the warehouse workers, even the people in the mailroom. If I didn't have a great team, I wouldn't be where I am today. Sure, I can get a few meetings and some investors on my name alone, but after they've got my autograph they want to see a plan and the product, so I have to perform. So far I've been successful at meshing sports and business. I always go to meetings dressed to play. There's no doubt that I'm prouder of what I do today than of what I did for the Bears thirty-five years ago. The key to my success is surrounding myself with good people. They know I'll be in at 5:00 AM and they can't beat me to work. They see I'm not late and so my people know not to be even a minute late to my meetings.

### DOES IT ALL MAKE SENSE TODAY?

Yes, it does. God gave me a gift and he took it away, and I'm sure it's all for the better. If I had played another three or four years, I wouldn't be in this business, nor would I have been the first black athletic director in the country.

### MY WRAP

*It's a rare person who could go from superstar runner to entry-level stockbroker. Sayers did it by using all the principles of sports to make him as respected in business as he was as a running back. He saw how good he could be and worked twice as hard as anyone else. So many are willing to just play on talent alone. Sayers had to max out his ability and is doing it in computers today as he did on the football field decades ago.*

# DARA TORRES

★ **8-TIME OLYMPIC MEDALIST: 4 GOLD—1984, 1992, 2000 (2), 4 BRONZE—1988, 2000 (3)**

★ **FIRST AMERICAN SWIMMER TO COMPETE IN 4 OLYMPICS, 1984, 1988, 1992, 2000**

★ **28 ALL-AMERICA HONORS AT THE UNIVERSITY OF FLORIDA**

Don't look back. Somethin' might be gainin' on you.

—SATCHEL PAIGE,
*Hall of Fame baseball pitcher*

I won because I hated to lose. There's no doubt I had natural talent, but I worked hard, and training consumed my life.

## THE MOMENT

*It's hard to believe, but after appearing in three Olympics, Dara's impact moment involved a comeback in 2000, after taking eight years off.*

It was 2000, and there I was at thirty-three trying to make a comeback in international swimming. Each day I went down to Stanford and trained with a group of elite swimmers in an effort to make the National team. My presence was evidently a turnoff to some of the other swimmers, and so I was kicked off the national team. I was shocked.

## HELP ARRIVES

Fortunately, Coach Richard Quick stepped up and offered to train me individually, which essentially kept my Olympic comeback hopes alive. The problem was that once the workouts started I was out there all alone, with no teammates to train with, to judge myself alongside of, to talk to. As a result, I just could not get motivated.

After a short time, I told Coach Quick that it just wasn't going to work out and he stopped me in my tracks by saying, "Dara, when you're out there at the Olympics are you going to worry about where everyone else is, or will you just be trying to do your best?"

Naturally, I said, "My best," and he said, "That's what practice is about—doing your best, not watching what everyone else is doing." I was sold and from January until the Olympic trials I trained by myself with Richard.

*She made the team and won five medals at Sydney, including two golds, tying Marion Jones for the most medals by an American athlete during those Olympics.*

I was able to feed off the negative energy, and it took me to my goal. Looking back, I should have realized how young my teammates were, and even though they accepted me they evidently didn't like how it was turning out and forced Richard to toss me. Looking back, if I had stayed on that team and did not get that push and the one-on-one training, I don't think I would have done as well or possibly even made the team.

## IT WASN'T ABOUT ME

The 2000 games stood out for me because of the number of people I felt I was racing for. So many women came up to me and said that they gave up on this or that, or that they were too old, or that their opportunity had flown by. But after seeing my comeback at thirty-three, after an eight-year layoff, they were inspired to try. I began to read these letters and hear these stories and think I had to do well, more for them than for me.

## ANOTHER TWIST

I said I have to win and hate to lose, and yet my most gratifying medals were the three bronzes I took in Sydney. Why? Because I had never won a medal in an individual event, so the bronzes were gold to me. In addition, my time in that race was as good as I'd ever done. Yes, I have to win at everything, but I knew I did my best and third was my best, so that was great.

## PERFORMING UNDER PRESSURE

I was always nervous before big meets and big races, but by 2000 I think I had it mastered. I would see these other swimmers looking so nervous and I thought, "I used to be that way." There was a time when before the big race I would feel sick to my stomach. But, thankfully, that changed, and by 2000 I would go through the process of asking myself a series of questions: Did I train the right way? Did I eat the right foods? Did I do everything my coach told me?

I answered yes to all those questions, and then I went out and actually had fun.

## MY WRAP

*You will not meet someone so accomplished yet less impressed with herself or what she did than Dara Torres. As far as she's concerned, that was then, this is now, and it's time to move on. In reality, she was and is a tireless worker who,*

*I think, would have been an All-American basketball or volleyball player because she had the perfect mix: immense talent and immense desire, a killer combo. This is a combination that worked for her and could work for you, too. Notice, too, that she had to win but could live with not winning. She stumbled but never fell. Her story is not about swimming and winning, it's about striving.*

# KURT BUSCH

★ NASCAR NEXTEL CUP CHAMPION, 2004

★ IROC CHAMPION, 2003

★ DWARF AND HOBBY LEAGUE CHAMPIONSHIPS, 1995–96

To achieve anything in this game, you must be prepared to dabble on the boundary of disaster.

—SIR STIRLING MOSS,
*legendary British Grand Prix winner*

They told me Little League baseball was supposed to be fun, but if I didn't win I was pretty upset. I guess you can say I've always had the competitive drive. I felt like I poured everything I had at that age into that game. When my team won, although I was satisfied, I was still looking to see what I could do better. To their credit, my family saw this will to win, this fire in me, and so my dad started to help me. When he couldn't, my uncle would come over and work with me on my mechanics. I had to be in the thick of things, so I usually hit leadoff or second and played catcher or shortstop.

## HOW BASEBALL MADE ME A BETTER DRIVER

I remember a particular big game that we had to win. It was tied in the last inning and our pitcher was beginning to struggle. He was getting wild, which wasn't good, because the other team had a runner on third. I was catching. Sure enough, after saving two balls in the dirt, the third squirted by me. As I went after it I said to myself, "There is no way I'm losing the game here." So I don't know how I did it, but facing the other way, I reached back, grabbed the ball, and with virtually no time to turn around, I hiked it like a football through my legs to the pitcher, who made the tag at the plate. We wound up winning in extra innings. That play taught me not only to never give up, but also that to win you sometimes have to be creative.

## A DAD'S SACRIFICE

We were and are a racing family. My dad raced all through my childhood, and when it started to get political and the costs started to rise, he sold his late-model car and bought two dwarf cars. As a result, he had to teach himself a brand-new way of racing, which was just another example of how my dad taught me how important sacrifice is.

We learned together, the hard way. He had me do anything the crew needed done, like washing the cars and handling the tire pressure. It was great.

At fifteen, I was finally going to get my chance to race, but first my dad and I built a car together. He wanted me to understand the mechanics of the car, and this was truly the only way. He didn't want me to go out there and tear it up without understanding what it would take to get ready for the next week's race. Being with him, side by side, building that car, was something I would not trade for anything. Even today, I take extra care not to tear up a car because I know how hard it is to get it on the track again.

### IF AT FIRST . . .

I took second in my first race ever and I thought I was ready to roll. My next three races I wrecked the cars, probably because I thought I could do this without any real training. My parents, on the other hand, after this run of wrecks, were thinking that maybe they'd made a mistake. I was thinking, of course, that I'd just had a run of bad luck.

### THE DIFFERENCE ONE PERSON CAN MAKE

I met this old-timer walking the track and he gave me the words that will last me a lifetime. He said, "Son, you are putting yourself in a position to wreck your car."

I said, "What do you mean? He just drove over my hood? How could that be my fault?"

He said, "Slow down, look ahead, and drive the other guy's car ahead of you so that you're not the jagged edge looking for a place to go."

As a result, I learned to study the field. That's what racing is about. It took someone I didn't know approaching me with that advice and about a year's worth of trial and error to finally get the formula right. I learned that there are times to use speed and times not to. Racing is more about strategy—picking your spots to use your speed. That philosophy launched my pro career.

### BECOMING A BETTER PERSON AFTER BECOMING A GREAT DRIVER

Right now, I'm focused on trying to become a better person in the eyes of the fans. I think I've been misunderstood, because all I've tried to do is be considered a "regular guy." I've learned that I have to work hard at being better with the fans and, at the same time, work hard to get better as a driver. My hope is that fans will realize that Kurt Busch is "a guy we can pull for because he was just like us at one point."

## HOW I PLAY THE GAME

I don't mind being the guy to beat, the guy who gets to the checkered flag first because he outthinks the rest of the pack. I think about the mental side of racing, the track, the restrictor plates, the field. I'm always trying to think ahead. That's what I take a lot of pride in. Over the years, I've found out that you can't hit a home run every time at bat and, to use a football analogy, there are times when you just need four yards more than you need a touchdown. You have to do what the team needs at that moment as opposed to what you think you need.

## MY WRAP

*Kurt Busch has gotten respect as a driver. Now he wants the popularity, and he's focused on making it happen. It seems to me that if the fans learn about his background, they will find a lot to cheer about. And keep in mind, Darrell Waltrip claims to have been booed twice as loud during his heyday, and he's now one of the most popular men in his sport.*

# MARTIN JACOBSON

- ★ 5 NEW YORK CITY TITLES IN THE LAST EIGHT YEARS

- ★ OVER 1,000 CAREER WINS AS A HIGH SCHOOL SOCCER COACH

- ★ OVERALL WINNING PERCENTAGE OF 93%

- ★ 30 YEARS COACHING AT MANHATTAN CENTER HIGH SCHOOL, LAGUARDIA HIGH SCHOOL, AND MARTIN LUTHER KING JR. HIGH SCHOOL

A coach isn't as smart as they say when he wins, or as stupid when he loses.

—DARRELL ROYAL,
*longtime University of Texas football coach*

I started playing soccer at ten and began coaching right out of college at Ball State in 1968 and just never stopped. I never saw color, race, or country, I just walked through the hallways of every high school I worked at, saw talent, and formed a team.

I took over a team at Martin Luther King Jr. High School in 1993 that was 0-12-1. We won the division in 1994. By 1996, we began a streak where we won eight championships and lost two finals in ten years.

### HOW DOES THIS HAPPEN?

The way I accomplished this was get into my players' hearts and souls. It never ended when practice was over. I had to get involved in their lives. Most of my kids had to work to support their families; many were new immigrants to this country and came here with nothing. Once I showed that I cared about my kids, they would bring their cousins and brothers in and a legacy was born. I'd help these kids take the test to get into the school and if they got in they had to show that it wasn't all about soccer. It was about the grades, the work, and meshing into the American culture. Eighty-five percent of my kids graduate, and the majority go on to college. I'm much prouder of that than any title we've ever won. It's at the point now where our team gets so much attention that our top player is usually one of the top national recruits. But I care as much about the twenty-fifth man on the roster as I care about the thirty-goal scorer.

### MY STYLE

I am as intense as Bobby Knight, but I'm mellow-intense. I'm not a screamer. I have a peaceful inside and it's probably because of my spiritual beliefs. I think the kids pick up on that and they trust me. Most of these kids are Muslims from Africa—you scream out the name "Muhammad" at practice and maybe nine kids will turn around—and yet they listen to a Jewish coach.

## IT'S NOT ABOUT THE MONEY

Jacobson is successful because he has no life except us.

—FORMER MLK PLAYER

Most high school coaches don't do it for the money—my first job was for $5,300. I'm not exactly living the high life now and I have three jobs: athletic director, guidance counselor, and coach. It's about passion, and it's my life. I often have these kids over for holidays and make sure they check in with me every day, even when we're not in season.

## JUST HAPPY TO GIVE BACK

This year, I'm twenty years clean from heroin and this gives me inspiration and a new zest for life. I'd been shot and stabbed while scoring drugs. Surviving that has changed my life and my coaching philosophy forever. This is my second chance and I do not plan on blowing it. There is a sign in my office that says, PLAYING SOCCER FOR MLK IS ABOUT GETTING A BETTER LIFE. And it's true. I really think it's about the heart, not the hardware. It's about the kids and the success stories, like a kid I coached from Mali who is now in medical school.

## FINAL THOUGHT

I would hope that someone describes me as a coach who cares about the kids, not a guy who wins championships. Drop by my basement office at 2:45 and you'll see that it's packed. If you listen, you'll hear very little talk about soccer. Instead, we talk about life. I listen to my kids' real-life struggles about trying to fit into this country and the hard times they left behind in their native lands.

## MY WRAP

*Stats don't lie. Coach Jacobson has had incredible on-field and off-field success with some of the toughest new immigrants to our country. As with many of us, Martin does not claim to have lived a perfect life, but like most of the people in this book, he's learned from his mistakes and has made an enormous impact on the lives of others. Martin has used sports to make that impact and he's not about to stop. I see a movie coming out of this!*

# BEN CRENSHAW

★ 19 PGA TOUR VICTORIES

★ RYDER CUP PLAYER, 1981, 1983, 1987, 1995; U.S. TEAM CAPTAIN, 1999

★ 2-TIME MASTERS CHAMPION, 1984, 1995

★ PGA TOUR GOLFER, 1973-PRESENT

Sometimes this game cannot be endured with a club in one's hand.

—BOBBY JONES,
*legendary golfer*

I got involved in golf at a very early age and I learn something new about it every day. I just love the history of it, and I often wonder why this sport has lasted five hundred years.

I am all about playing by the rules. You play and live your life by your honor. You play hard and you play competitive, and then, when it's over, you shake hands and, if you've lost, you accept that with all the grace you have.

### "OH, BENNY"

My battle is with my temper, and it still bubbles up every so often. I've thrown clubs and I broke a driver one time in front of my mother in a college match. She was so embarrassed and ashamed, it was terrible. I still regret it to this day. I can see her sad brown eyes and her saying, "Oh, Benny." It was tough! As a pro, I threw a club at my own bag and got fined. To this day, before I lose it on the course, I can hear my mom, and most of the time that stops me from doing something I'll regret.

### MEETING A LEGEND

I happened to meet a legend growing up named Harvey Penick. Many people talk about how he taught me to putt perhaps better than anyone else on the tour, but I think about what he taught me about life. I met him was I was very young and I'm grateful for all he gave me in my life. I'll give you an example. I was ten years old and one guy was driving away on the range when a young Mexican man started collecting the balls on the range. Well, the guy who was hitting started screaming at the worker to get out of his way. That's when Harvey flew over to him and said, "Look, that young man is just doing his job. He's as important to this club as anyone, so why don't you hit a different direction." I thought, *"Wow, he cares about what's right, not about who has the most money or power."*

### DEFINING MOMENTS ON THE TOUR: THE MASTERS

Harvey passed away on April 2, 1995—I served as a pallbearer at the funeral a few days later—and the next day I played in the Masters. I won that tournament for him and always felt fortunate that I was able to salute my mentor with something that signified everything he taught me. In his last days, he was still helping with my game, in particular my putting. Some say, "Why play golf in your last days?" But it wasn't about golf. It was about how Harvey had dedicated his life to helping other people. He was selfless to the end, and it's something I take with me every day.

## UP AND DOWN

I still can't figure out why I sometimes feel incapable of making a mistake and at other times I'm flat-out fragile. It's the way both life and golf are. I wish I could tell you I've solved this issue, but I haven't . . . yet. I still relish those pressure moments, win or lose. I want to be in it at the end, which means here comes the pressure. I love it. My first Masters win was a relief because I had been close to a major championship eight times and had come up short on every one of them. At thirty-two, I won it and right then I began to believe I was capable of being a great golfer. Up until then, I had some doubts. But there's nothing like success to prove to yourself you can do it.

## FINAL THOUGHT

Looking back at my life, I realize I learned to respect rules in golf, which, in turn, taught me to respect the rules of life. I've learned from the sport to act honorably, even when no one else is looking. And I like to think that's the way I've led my life. I will always strive to do so. I guess you can say I learned that first life lesson on the green.

## THE WRAP

*Ben Crenshaw defines the word "grateful." Yes, we know he's talented, and he thinks of himself as fortunate. Again, we see the power of one. And in this case, that one person was a pretty famous one—Harvey Penick—who cared enough to help a kid and infuse him not only with golf technique, but with life's lessons.*

# DANNY AIELLO

★ NOMINATED FOR AN ACADEMY AWARD FOR BEST ACTOR IN A
  SUPPORTING ROLE FOR *DO THE RIGHT THING*

★ FEATURED IN SUCH MOVIES AS *ONCE UPON A TIME IN AMERICA,*
  *THE GODFATHER PART II, THE PURPLE ROSE OF CAIRO,*
  *MOONSTRUCK*

★ NOTED BROADWAY AND FILM ACTOR

Whoever would understand the heart and mind of America
had better learn baseball.

—JACQUES BARZUN,

*writer/historian*

Baseball played a major role in my life both as a player and a fan. As a kid, I would get to the major league games as early as possible, not to see the hitting but the fielding practice, because the perfection they attained was so unbelievable.

I was a switch-hitting, long ball hitter, a Dave Kingman type. They called me "Three Sewers" Danny in the Broadway Show League. I hit these long, high balls. I was a dead pull hitter, so I never had a great average. But I could hit, and could hit it far. I never was paid to play, but I did garner some interest from the majors. There was a scout after me from the New York Giants. He came to my house and talked to my mother. They wanted to see what kind of kid I was and what kind of family I had. But I didn't think too much of it, because I was just a kid. The St. Louis Browns also sent a scout to look at me play in the Federation League. I don't know anyone who was ever signed by these guys, but they did come to many of the games in Manhattan and the Bronx.

## OWED TO BASEBALL

I also played every day in the army, in Munich, Germany, for three years from 1952 to 1954. It got me out of so many marches and drills. I would be walking out onto the field with my mitt, passing my unit on forced drills. It was incredible. If I didn't know how to play the game well, I would have been doing those drills and not been out there on the field every day. We even had a couple of major leaguers on our team, and that was a great way to judge my talent level.

## YOU MOVE LIKE AN ATHLETE

Playing sports helped me understand my body and taught me self-awareness, which, as an actor, is a huge asset. I was told I moved, walked, ran, and acted like an athlete, and that was a huge compliment. If I have to throw a punch in a movie, it has to be a gorgeous punch. I want people to know I can hit like a son of a b. When I played first base in *Bang the Drum Slowly*, I had to look like a major leaguer. I had to stretch for a ball like a major leaguer, my feet had to be right, my hand positioning had to be dead on. Bobby Murcer (*a former Yankee, Giant, and Cub*) said, "I saw you switching feet." The real players who appeared as extras were ripping Bobby De Niro (*Robert to you and me*) as was the worst baseball player they had ever seen.

### I CAN DO THAT. JUST LET ME TRY

I was a bouncer at the Improv and this playwright came up to me and said, "Do you want to act?"

I said, "Sure, why not?"

He gave me a fifteen-minute monologue to see if I could do it. It was a test and I knew it, and I just would not permit myself to fail. I grabbed that script and focused on it as if I were at home plate. Nothing else mattered. I absorbed it. I was used to the pressure of playing ball when, at some point, the game was on my shoulders. Having dealt with that kind of pressure lined me up for success onstage. Like acting, baseball is a game of concentration and relaxation. You've got to relax and not be afraid at the plate. Most of the guys who do not have success in baseball are the ones who think they are going to get cracked in the head with a ninety-mile-an-hour fastball. They never become great hitters. As an actor, if you can eliminate the fear, concentrate, and relax, you can do wonders.

### WHAT MIGHT HAVE BEEN

The timing was never right for me to give pro baseball a shot. I left school early, because I got into trouble, and went into the army. If the scouts were still interested they wouldn't have been able to find me. I got back from the army at twenty-one and just kept playing Federation ball all around New York until I was thirty-five. I played on a team called the Bronx Braves. I never got paid, but the quality of the games was tremendous. To this day, I say I could have been a major leaguer. I was not a fast runner and I didn't use the whole field when I hit, but if I played every day and got the proper coaching, I think I would have had a real shot, at least playing in the minor leagues.

## STICKING IT OUT

I'm the kind of person who always perseveres. I never quit baseball, I stayed in it for decades. And when I dove into acting, I was and still am the same way. I work whenever I can and always look to get better. Even today, I just don't quit.

## NO REGRETS

If I hadn't gotten into gang fights, spent so much time in pool halls, and stayed in school, maybe I could have made it to the majors. But I'm not the kind of person who looks back. The more I think of it, the fact that I didn't give myself a chance to make it as a ballplayer made me more determined to not shoot myself in the foot as an actor. The dedication and perseverance, not taking no for an answer, and not letting myself get distracted are what I learned had to be done to be successful as an actor. I made it as an actor and never knew a soul. I just said to myself that I would not fail. I had two kids, was making one hundred fifty dollars a week as a bouncer at the age of thirty-six, and still found a way to break through in acting. Now I have done over eighty movies and turned myself into a singer and I'm working on my second album. If I had done this in baseball, God knows how far I could have gone.

## MY WRAP

*Like many of us, early decisions helped keep Danny from maxing out his ability in sports, but it did set him up for success. Again, we see a negative situation turned positive because the person chose to learn from rather then dwell on his failures. I think I'm going to go out and rent* Bang the Drum Slowly *and buy his album. How about you?*

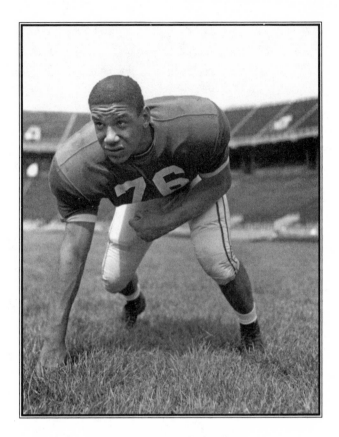

# JIM MARSHALL

- ★ INDUCTED INTO NFL HALL OF FAME
- ★ PLAYED IN 2 PRO BOWLS, 4 SUPER BOWLS, RECOVERED AN NFL RECORD 29 FUMBLES
- ★ PLAYED IN 270 CONSECUTIVE STARTS AT DEFENSIVE END, AN NFL RECORD
- ★ NFL DEFENSIVE END, CLEVELAND BROWNS AND MINNESOTA VIKINGS, 1960–1979

Football is blocking and tackling. Everything else is mythology.

—VINCE LOMBARDI,
*legendary Green Bay Packers football coach*

Sports reinforced everything I learned from my father and grandfather, and that was all about having a good work ethic. They used to say, "Be ready to sign your name to every job you tackle, because people will examine it and you have to be able to say you completed it with pride." That was my approach to sports: give your best and be a teammate anyone would consider dependable.

*I'd say that playing every game for twenty years could be considered dependable, don't you?*

## THE MOMENT THAT CONVINCED ME I COULD DO IT

I was in high school and a runner was coming down the sidelines. A blocker came in on me and I jumped over him and made the tackle on the sidelines. One of the coaches, who it turned out had been a great player in his day, came up to me and said that I could be special. That meant a lot to me, and I started to believe that I was a special player and I began to play that way. I had always tried to do my best, and I was brought up to not care about the acclaim but just to live up to my own expectations, and that conversation with my coach raised my expectations considerably.

## ALL ABOUT DOING IT RIGHT

I remember walking with my dad one day and he pointed out the roof of a huge house. He asked me what I saw and I said, "Two guys replacing roof shingles."

He said, "Notice how they're placed. Very carefully, so that the water passes from one to the next until the water runs off."

It took me a while to get his point, but we watched for hours as these two roofers place one tile after another across the roof. I finally got that he was trying to teach me the need for teamwork, and the importance of doing a job the right way. If one tile was out of place that roof wouldn't drain properly and eventually it would leak and rot. It's all about integrity and the need to do things right, if you want things to work properly. You have to seek a degree of perfection, and my dad and grandfather were just drilling it into me. I was taught never to accept second best.

*Perhaps that's why Jim Marshall played longer and more consistently then any defensive lineman in NFL history. "I did what I was suppose to do," is the way he'd put it.*

## NOBODY'S PERFECT

Striving to never be second best is what makes the four Super Bowl losses so tough. The fourth, versus Oakland, was the only game I ever played in that where I did not make a tackle. I fell into a state of depression. I tried as hard as I could to be good, but that day I just wasn't close to my best effort, and it was tough to deal with. It happens in life and now that I'm retired, I've come to a resolution with this and I'm no longer ashamed.

## THE WRONG-WAY RUN, OCTOBER 25, 1964

One of the biggest mistakes I ever made was when I picked up a fumble and ran the wrong way, sixty-six yards, into the wrong end zone.

*Yes, his own team's end zone. It's never been done since, and just about all football fans remember it.*

In truth, I had one of my better games that day. Ultimately I made up for it by causing a fumble, which my teammate, Carl Eller, took in for a touchdown for the winning score. What did I learn? To think things through! That's what. What I am proud of is that I got myself together and, in reflection, it was a good character-builder. I found that people relate to the play because we all make mistakes in our life. It just so happens that I have to answer for that play, for that mistake, every day since, which isn't usually what happens with other people.

## IT'S ALL ABOUT GETTING UP, NOT GETTING KNOCKED DOWN

I look at life as a game, and the game of life is the most difficult game we can play. We are faced with so many different questions and decisions every day, and sports helps us set up a framework to deal with all these issues. My dad's lessons have given me my own rules and guidelines that have helped me and will continue to help me. All I know is that I play life and football to the best of my ability and more often then not, it's more than enough.

## MY WRAP

*It seems unjust that a person who exhibits such pride and consistency his whole life would be linked with five negative moments—the wrong-way run and four Super Bowl defeats—but it says so much to me that he has put it behind him. Jim Marshall knows what we all should know: that no one is perfect, and that we can all control our effort, but not necessarily the result.*

# BRANDI CHASTAIN

★ **COVER OF** *TIME, NEWSWEEK,* **AND** *SPORTS ILLUSTRATED*

★ **SCORED WINNING GOAL OF 1999 WOMEN'S WORLD CUP FINAL ON PENALTY KICK**

★ **GOLD MEDAL, U.S. WOMEN'S OLYMPIC SOCCER TEAM, 1996**

I was trying to give the league a little exposure.

—PAUL CANNELL,

*Washington Diplomats player, explaining why he dropped his*
*shorts during a North American Soccer League game*

Two defining moments for me came years apart. One was in 1980, when my junior high soccer coach, a male, told me I could not play because I was a girl. All the boys around me asked, "Why not? She's as good as the rest of us." Up to that point, I'd never been told no in my life. When I realized I had teammates who liked the way I played well enough to back me, I realized my play on the field could actually win people over. I played on that junior high team, put up with our opponents' disbelief, and always had my teammates in my corner. It was a great season for me because of that.

My second moment happened playing college soccer. I missed two and a half years with knee injuries and I ended up switching schools, to Santa Clara, where the physical demands were like nothing I had ever seen before. When the running became too difficult, I just quit. I said my knee was bothering me and that I had to sit out. In my head, I thought, *I am so good, I don't need to do all that. That's for the other players who aren't at my level.* I knew that when the game started, I would score the goals if they'd just get me the ball. One day my coach—and future husband—pulled me into the office and said, "This is ridiculous. You're setting a horrible example and you need to do the work or leave the team." So I quit. Just like that, I quit the team. Rather than do the work the team was doing, I walked away. Well, in the next forty-eight hours I came to realize I had made the mistake of my life. I went to the office and asked for another chance. Thankfully, he gave in. He scheduled an entire session of fitness. I came in first on every run and I learned my lesson.

### SHE OWES IT ALL TO DAD

I owe my dad for my approach to the game, because he was my first coach. He was a marine and we learned soccer together from a book. Going to the ball, showing up on time, giving maximum effort, that's what he always demanded. He would send me home a lot for not paying attention or fooling around. In the end, my years with him on the field gave me the mental and physical toughness to excel later on.

## WILLINGNESS TO CHANGE

Overall, change has been the best thing to happen to me in my life: the transformation as a player at Santa Clara; the switch from being single to being married; going from just married to becoming a mom; and on the national team, switching from forward to defender. Each time it happened it was scary, but each time it worked out for the better. The key is sticking it out in the beginning of change, because at first it always looks bleak.

## NATIONAL TEAM

In 1991, I did not play very much. From 1992–94, I was rarely with the team. I went to Japan to play pro and it was great, but in the end, going pro cost me the 1995 World Cup, because I was not in the mix. When Tony DiCicco took over, I picked up the phone and made a call, but I still missed that World Cup. When the player strike happened the next year, I was invited into camp. At the end of camp, the coaches pulled me over and let me know that I had won them over. Then they said, "We need you not as a forward but as a defender." My jaw dropped, but I was smart enough to know this was my big chance, so I embraced it. The rest is history. I started and played every game up until the 1996 Olympics.

## THE BRA

> If you look at mistakes as burdens, then you'll never have the opportunity to get better.

That penalty kick and what I did after it have been great for U.S. soccer and of course, for me. But it was just one moment. There were others. Remember, I scored in my own goal against Germany, having Carla Overbeck say, "Don't worry, we'll get it back." Later, I would score to tie the game, 2–2. There was Mia Hamm scoring late to beat the Czech Republic; Kristine Lilly clearing the ball off the line versus China. All these moments showed me the meaning of team and they were all special to me and everyone on the team. You and others may think of the "bra," but the team knows it was about them not me.

*Okay, so you didn't see the cover of* Time *magazine and* Sports Illustrated *in July 1999, when Brandi scored the winning penalty kick to give the United States the Women's World Cup in front of one hundred thousand at the Rose Bowl in Pasadena, California.. Well, it was the way she celebrated and not how she scored that stood out. She took her shirt off, exposing her sports bra. The sports world had seen male players do that, but for the women, well, this was a first. Her life and the game have never been the same.*

### FINAL THOUGHT

Everything I learned about life, about change, about attitude, I learned from soccer. As a stepmother of a seventeen-year-old, and as a wife and new mom, I often go back to soccer to think about what I learned and how to apply it. Everything I learned on the field works for me off it. I always look at people in this world and think we are in this together. It's why I might give some insight or tips to opponents on the field as we play. I like competing, but I do not hate my opponent.

### MY WRAP

*Brandi may be the most famous player on that championship team, but she may have had the hardest road of all the core stars on that national team. You get the feeling she'll do as well in broadcasting as she did as a player. Her journey to the top makes her even more identifiable to the masses than most megastars, and that's probably why kids especially can't get enough of her.*

# LARRY FINE

Boxing is sort of like jazz. The better it is, the less amount of people can appreciate it.

—GEORGE FOREMAN,
*two-time heavyweight boxing champ*

Growing up, I lived right near my grandfather. As his oldest grandson, I was able to spend a great deal of my youth by his side. In fact, I was even able to work with the Stooges on the set.

Larry boxed at a pretty high level for about five years when he was young. He had over forty fights and won most of them. He was a small guy, about 5'4", but the strength of his hands and wrists was unbelievable. I knew this because he would spar with my dad and me all the time and till the end his strength was incredible. As a fighter, they say he was a guy who was willing to take the punch. He became a counterpuncher and relied on his quickness to do the damage.

He boxed to earn money, because he decided in his teens he wanted to go into show business. He used his earnings from boxing to learn the craft of acting, which included taking up the violin. He also considered boxing part of the entertainment world. Simply put, he fought to keep the money coming in after World War I.

Even after he stopped boxing, Larry was a fervent fight fan. I would go with him to the fights each week in Los Angeles, and we'd often pick up his fellow Stooge Moe's brother, Shemp, who, by the way, died of a heart attack in a cab after a night attending the fights.

Until he hit middle age, Larry stayed in great shape and claimed his days boxing helped him quite a bit in show business.. The footwork, the eye–hand coordination, the athleticism, were all evident in the Stooges' shows. He applied the same work ethic of philosophy in all his performances. We used to watch Moe and Larry practice all their physical scenes over and over again. I used to say, "Why do you guys go over this so much?"

Larry answered, "We're good because we *do* practice like this, and it's also the reason we don't get hurt." He used to say, "Boxing taught me to stay in shape and how to maintain a physical edge."

Up until their late thirties, all the Stooges were really chiseled out of stone.

## FEAR

Having been hit in the ring early in life, Larry eliminated the fear of being hurt in any physical activity. That attitude was transferred to the movie set, because he never worried about being hit or getting hurt when performing. The hitting was an illusion for the most part, but even when it wasn't, it was okay. He also told me that he knew how to take a punch, how to make it look real and go with it, especially at those times when the fake hits accidentally became real hits. Also, the balance needed to do those stunts was incredible, and he credited the fact that he had that balance to his time in the ring.

## ON THE IMPACT OF BEING BOOED

Larry told me that if he hadn't boxed he didn't know if he could have performed the physical stuff. It gave him the confidence that he could handle all the physical routines he performed with Moe, Curly, and Shemp. Vaudeville demanded three shows a day, bad conditions, and lots of travel. It was a grind, but his days in the ring steeled him for his early career, especially when he was booed by fans.

## MY WRAP

*In the early days, boxing sustained Larry financially and kept paying off for him throughout his life. Larry knew he'd never be a champ, but he approached the sport with the same passion as he did his acting. You don't know how any sport will help you, but for Larry Fine it helped him develop his balance, fitness, and overall physicality. But the unexpected bonus he may have been surprised he'd acquired, according to his grandson, at least, was the confidence to attack life and a fearlessness about the possibility of failure.*

*Beebe is wearing number 82*

# DON BEEBE

★ **HOLDS THE RECORD FOR SUPER BOWL APPEARANCES BY ANY ONE PLAYER: 6**

★ **SUPER BOWL CHAMPION, 1997**

★ **MEMBER OF 4 BUFFALO BILLS SUPER BOWL TEAMS, 1990–1993**

★ **WIDE RECEIVER, BUFFALO BILLS, GREEN BAY PACKERS, 1989–1997**

He's such a good person, it gets overlooked what a good football player he is.

—BRETT FAVRE

I remember, when I was seven years old, asking God if he'd help me be something special in sports. This doesn't surprise me now, since I was so competitive even at such a young age.

I played so many sports well, but believe it or not, basketball was my passion. As a matter of fact, if it wasn't for my dad I wouldn't even have gone out for football. I asked him if I could just pass on football my junior year so I could throw myself into basketball. He said, "Run cross-country or play football. You make the call." I chose football.

*Good move. Attention football parents: Don Beebe did not play football until high school and yet he made it all the way to the top.*

### WHY I PLAYED AND WHY I NEVER QUIT

I played sports every season for as long as I can remember because of my dad. I didn't play because he did, I played because he didn't. Growing up, he was begged to go out for track and football because everyone knew how fast he was. But he said no. Instead, he hung out and smoked cigarettes and to this day, every day, he regrets it. He let me know I would be playing and that once I started, I would not quit.

### THE DIFFERENCE ONE PERSON CAN MAKE

Coach Joe Thorgesen instilled the passion in the game that I still have today. I went from a guy who had to play in the NBA to a guy who was playing in the NFL. He told me I could be great if I applied myself, and he did it in such a positive way it just resonated. He's also the reason I am a high school coach today. I see so many kids getting beat down emotionally by their moms, dads, and coaches that if I can help in any small way by instilling passion in the game and helping kids put it all in perspective, that's what it's all about.

### NOT FITTING IN

Junior year, I was expected to figure in much of the offense. Coach Thorgesen called me aside one day and told me that the seniors wanted nothing to do with me, that they had a maturity issue and that I should stay the course and that things would fall into place. And they did. I accepted that I would not be one of the guys, and so I dealt with it. My senior year ended up great, and I'm just so glad my coach spoke to me and that I stuck it out. However, if I see something like that happening with the teams I coach today, I crack down right away.

## DID I SAY, "NEVER QUIT?"

I had a full-ride football scholarship to a Division I school, which from the small town I grew up in didn't happen often. For a 5'10" white guy, that does not happen much.

*Wait a minute, I'm a 5'10" white guy.*

So I go to preseason football camp at Western Illinois and I got so sick. I couldn't eat, and I was throwing up. I lost twenty-one pounds. It was horrible. I called my dad and said I couldn't stay there. It was a combination of missing my girlfriend, who I wound up marrying; being homesick; and being ill. I ended up leaving. For three years I stuck close to home, working in construction, playing recreational basketball. Then something happened. I got football fever. I became obsessed and actually tried out for the Bears. Yes, the Bears! Essentially, they told me to go home. I did, and I enrolled at Western Illinois, but they couldn't make me eligible to play. I wouldn't give up. I enrolled at Western Illinois again and played a year.

I wasn't ready for the NFL yet, so I transferred to a small NAIA [National Association of Intercollegiate Athletics] school, Chadron State College in Nebraska. (Playing for a NAIA school was like playing at a NCAA Division II or III level.) Most Division I athletes didn't go on to play pro ball and the odds were considerably higher to even get a look from a NFL team if you played for NAIA school. Well, I beat the odds and got myself invited to the NFL Combine. I played well there and had some interest, but it was nothing serious. Through some fluke, a scout timed me in the forty-yard dash and word got out that I could run. I got into the NFL combine and suddenly a lot more people seemed curious. There I ran the forty and tied with Deion Sanders as the fastest ever. Suddenly the Packers and Raiders were at my doorstep. The Jets called, too, but it was the Bills who drafted me.

*You can't make up a story like this, can you? But it's all true! It's likely you've witnessed the rest of his story. Oh, and I did mention that you should never quit?*

## THE PLAY DEFINES THE SPORT

The scene is Super Bowl XXVII. We were losing to Dallas 52–17 and Frank Reich fumbled the ball. I was running a fly pattern down the left-hand side when Leon Lett of Dallas scooped up the ball. I turned around and started running after this 300-pound lineman. Somehow, I caught him, but I couldn't tackle him. He was just too big. But when he chose to put the ball out by his side, I knocked it away. When I did, he kneed me in the head and let me tell you, I was ticked. I didn't think much of the play and I didn't know what was happening across the country or in the television booth or anywhere else for that matter. I just knew we got blown out. After the game, I was sitting by my locker when the owner, Ralph Wilson, came up to me with tears in his eyes. He said, "That play represents what we are about in this organization."

Later, when I got to the podium, it was all people wanted to talk about. Not that we got whupped. They wanted to talk about Leon Lett and that play. I got boxes of fan mail letters, blowing away by millions the number of letters Jim Kelly got. Many were from dads all across the country. One said, "I never had a great relationship with my son, but after that play I told him this is what it's all about and we talked for hours. Our relationship has changed. We're tighter now and we understand each other better."

For me, it was only one football play, but for a good part of the country it meant so much more. What most people don't realize about the play is that it wasn't just about me running him down, it was about him not hustling into the end zone.

### MY WRAP

*The late Dick Schaap once said, "Don is one of the greatest stories in sports."*

*He was right. Don Beebe's whole life led to this particular play. Beebe almost never quit, and if there was one guy who could catch Lett and would be driven to catch Lett, it's Don. This story is the stuff of legend. It shows the heart of a man. It personifies why I wrote this book. It's all about character, and that's why it's so perfect that after losing four Super Bowls he finally won one in Green Bay.*

*The bottom-line lesson: if you try your best, you'll always be able to live with the results.*

*Colangelo, center, with ball*

# JERRY COLANGELO

★ **ELECTED TO NBA HALL OF FAME, 2004**

★ **FORMER OWNER, PHOENIX SUNS (NBA), ARIZONA DIAMONDBACKS (MLB), PHOENIX MERCURY (WNBA), AND ARIZONA RATTLERS (AFL)**

★ **FORMER COACH AND GENERAL MANAGER, PHOENIX SUNS**

When you are not practicing, remember, someone somewhere is practicing, and when you meet him he will win.

—"EASY" ED MACAULEY,
*Hall of Fame basketball player*

E ven though I was more of a baseball player, I tried out for the basket-ball team in junior high school. I got cut. What made matters worse was that the coach, Jim Bogan, embarrassed me in front of everyone when he said, "Kid, you better learn to shoot a layup with your right hand before you try out for my team again." (I was a lefty.) That moment had such an impact, his comment cut so deep, that I went right to the playground to practice that right hand. I was intent on making his team in the eighth grade. Not for a second did I think about quitting.

I learned how to make that layup, made the team, and by the middle of the season I was a starter. In high school, everything came together, and by the time I was a senior I had sixty-six college scholarship offers. All this led to forty years in basketball and an induction into the Basketball Hall of Fame in 2004.

**WHAT I LEARNED**

From that experience, I learned that I had to be prepared for everything. I deserved to be cut in seventh grade, because my game was not ready. What I did from then on was make myself ready for anything. I used to eat raw carrots to better my eyesight. I would shoot relentlessly at night, without using the backboard or a net. I thought that if I could just hit the rim under those conditions I would be a better player.

I owe this attitude to the coach who brazenly cut me years ago. I resented him at the time, of course, but his review of my game motivated me to the extent that I had an extremely satisfying career as a player, scout, coach, general manager, and owner. I just had to prove him to him I could do it, and I still keep in touch with him to this day.

*Amazing story! But why didn't you quit in seventh grade?*

Growing up in the neighborhood I did, in Chicago, set me up to react like I did when I was cut. The neighborhood was full of tough, fair people who worked in factories and steel mills. They didn't have very much, but what they had they shared. It's where I learned about family commitment, hard work, and passion. Growing up in this environment meant I didn't quit anything easily, let alone because I couldn't hit an offhanded layup.

## THRIVING UNDER PRESSURE WITH AND WITHOUT THE BALL

In 1987, I was working to buy the Phoenix Suns for $44 million, which at that point was the highest price ever paid for an NBA franchise. In 2004, I sold the Suns for $400 million.

*I guess you could say it was a solid investment.*

I had six weeks to put a deal together, and at the last minute I had two corporate partners back out. I remember my lawyers saying, "Nice try. You did all you could do, but this deal is dead."

I said, "No, it's not." I stayed up all night and hatched out a plan. Suffice it to say that it put a lot of the financial pressure on me and demanded a lot of trust from the sellers. I offered the new plan to the seller, and we closed, to the surprise of just about everyone I knew. We closed on a Friday and that Monday was "Black Monday" on the stock market, which meant that the deal never would have happened had I not closed the day I did. That purchase led to building a new arena that allowed hockey to come to Arizona, which then led to Arizona being awarded a baseball franchise. I think the composure and ability to think creatively under pressure come from my days competing on the court.

## WINNING

I hate to say it, but my greatest lessons in life were learned after losses. For example, when I was a high school senior, my whole ambition in life was to win the state championship. After leading the entire game, we were beaten in the final seconds by one point, as the last shot rimmed out. I thought my life ended because it was all that mattered to me. The funny thing is, the sun did come up the next day. Soon after, I reset my goals and moved on. Losing is hard, but somebody has to win and somebody has to lose. Sometimes just being in the game is the thrill.

**MY WRAP**

*Once again a Hall of Famer describes the hard times, which lets us know that they happen to everyone. It's not whether you win or lose, it's what you do next and how you act that defines what you will do with your future. Things have worked out pretty well for Jerry Colangelo because he never stopped hustling for the loose ball in life or in business. Look out for his son, Bryan, now who is general manager of the Toronto Raptors.*

# ABRAHAM LINCOLN

★ 16TH PRESIDENT OF THE UNITED STATES, 1860–1865

★ U.S. HOUSE OF REPRESENTATIVES, 1847–49

★ ILLINOIS HOUSE OF REPRESENTATIVES, 1834–42

The taste of defeat has a richness of experience all its own.
—BILL BRADLEY,
*former NBA basketball player and U.S. senator*

INTERVIEW WITH DOUG WEAD,
AUTHOR OF *RAISING OF A PRESIDENT*:

In the frontier, wrestling was very common. In fact, Lincoln's mother had been a wrestler. She was such a good wrestler that they used to place sucker bets for drinks with strangers who were passing through. They'd say, "I bet you can't even take one of our women," and then they'd pair them up with Abe's mom, Nancy Hanks. She'd often win and make some money for the family.

As for Lincoln, he was very strong, probably because his father put an axe in his hands at the age of eight and he swung that axe his whole life. As you know, he was tall, lanky, and sinewy, and he had strength that didn't show. He didn't look like a bodybuilder, but he was extremely strong.

His father, Tom Lincoln, was thrilled with his son's strength, and his dad would set up matches for him. But Abe would irritate his dad by shaking hands before and after a fight and wouldn't finish these guys off like he should have.

## ONE MATCH SAYS IT ALL

> Do I not destroy my enemies when I make them my friends?

When the local champion, Jack Armstrong, began hearing stories of Tom's boy, he came to town and challenged Abe. His father accepted the challenge for his son. From the start of the bout, Lincoln thrashed the local champion. Armstrong, who was frustrated by Lincoln's lanky build and enormous reach, started fouling his opponent. Abe put up with it for a while, and eventually he won. After the bout was over, he made sure to shake Armstrong's hand, much to the fury of his dad, who thought that showed weakness. The crowd was furious, and it looked like the mob would erupt. It was Armstrong who made sure the mob didn't put a beating on Lincoln, and he and Abe ended up being lifelong friends.

**LESSON LEARNED**

When Lincoln joined the militia briefly to combat the Indians, he had two fights and lost both of them. He was surprised at his losses, but concluded thus: "No matter how good and big you think you are there is always someone better and stronger."

**MY WRAP**

*Lincoln was tough, strong, and respectful, and his dad could not have been more dissatisfied. What he didn't learn from his abusive father and his stepmother he learned working the land, reading, and, believe it or not, fighting. His tough upbringing and gritty approach to challenges helped him lead a nation through its most challenging time. If Lincoln could only have gotten a hand on John Wilkes Booth, he might have been able to see the end of that play.*

# COACH VINCENT O'CONNOR

★ **9 PLAYERS HAVE GONE TO THE NFL**

★ **17 LEAGUE TITLES**

★ **20-TIME WINNER OF COACH OF THE YEAR**

★ **300 WINS AND COUNTING**

★ **53 YEARS COACHING NEW YORK CITY HIGH SCHOOL FOOTBALL**

Coaching is an unnatural way of life. It's natural for me because I'm used to it, but it would be unnatural for most persons. Your victories and losses are too clear-cut.

—TOMMY PROTHRO,
*college and NFL football coach*

My philosophy evolved over the years, and during that time I started to understand all about being unselfish. We're talking about football here, but, of course, what we're really talking about is values for life. The locker room is really a microcosm of the world, and it's important to me to have unselfish teams. When I don't, I'm not doing my job. I tell the kids I work with that they must think of their teammates and about the other team.

## MY WAY

> He teaches you football at the same time he teaches you about life.
>
> —JEFFREY AIME
> *team captain, 2005*

I never differentiated between the superstar and the sub. Let's take Dan Henning, a longtime NFL head coach and coordinator now with the Panthers of Carolina. He was the best quarterback we ever had, and that's why he's made it as a coach for twenty-eight years. He could analyze anything, even as a kid. It's a different story with Bill Pickell, who was a late bloomer as a player, but did he ever bloom. He did well at Rutgers, but he really hit his stride in the NFL with the Jets and Raiders. Pickell was not starting until late in his senior year because he didn't grow into his body in high school. Rutgers was interested but not interested enough to offer him a scholarship. So I pushed Billy into prep school. I knew by the time he was done he'd have grown into his body and, sure enough, after a year, Rutgers was interested and away he went. This has happened countless times. I'm not saying everyone I've coached has gone to the NFL, but I was able to help place a lot of players who were not stars.

## MY GAME

Throughout my childhood, I felt invisible. The first time a teacher or anyone really knew me was in high school. It wasn't a teacher but a coach who singled me out. Don't get me wrong, I'm not saying I was a good player. I was a backup fullback, I kicked and punted, but I was on the team and that was special to me. As the child of first-generation immigrants who were caught up in the struggle to survive, I didn't get much guidance or recognition from my parents, not that I could blame them. The push was for me to get a job as a civil servant, not scoop up a coaching position. My parents were thinking security and I wanted to be fulfilled. Being fulfilled did not include the civil service.

My coach was not a yeller. He talked about the importance of being on the team, and he was always talking about life. I knew I wanted to coach and I wanted to latch on to his approach. My problem was that I wanted to play in college, but at one hundred seventy-five pounds with modest ability, I did not have options. So I worked in a factory and played sandlot ball.

*So what did the coach learn as a player? To take the initiative. To find a way not to wallow in self-pity or run with the wrong crowd.*

## THE MISSION AND THE MOMENT

I knew what it was like trying to find a college where I could play and go to school because I was immersed in that struggle, as were so many of my teammates. We were not standouts, but with the proper encouragement and direction, who knows what we could have done. I ended up getting drafted into the service and then using the GI Bill to go to college. I went to NYU and one day I walked over to St. Francis Prep and asked for a job. They gave me a job coaching the junior varsity. I started coaching at two hundred dollars a year, but the important thing was, I had a job.

I loved it from the beginning, from the camaraderie with the coaches to mentoring the players. Then the real work started. I became a student of the game, and in order to help the kids I knew I had to have success. I didn't play college ball, so I had to establish credibility somehow. I did it by pure work and study. We won, and suddenly the kids were listening.

*They did listen and they still must be listening, what with over three hundred wins over fifty-plus seasons.*

## MY FOCUS

Most kids do not believe in themselves, so the first thing I want to give them is confidence. I make them believe they can do what I ask them to do, and as a result, they can compete against the kid who is bigger and stronger, if they do what I say and play as a team. To have success, I became a fundamentalist, starting each season with the ancient T formation. Sure, the kids wanted a pro set, but I wanted them to learn pure, basic football. That took the pressure off the quarterback. We never had the athletes, but we won with deception and execution. I taught the kids to practice like they should play. I worked for them to get into the best school possible, where they could play and study if that's what they wanted. I was able to personalize coaching, to let them know I cared, because I did. I let them know I felt that they mattered. I found out about their families, learning something about them so they would lay it on the line for me when the time came.

## MY WRAP

*When most fans think of football coaches they picture men like Bill Parcells, Vince Lombardi, or Jimmy Johnson. But when you talk to most players, the coaches who make the major impact are the high school coaches. Coach O'Connor is one of those coaches. Please take note—he is self-made. He admits that his tactics and techniques evolved over time and that he grew in the job, as he had been far from a finished product when he arrived. Remember the one line earlier in the story when, as a kid, he felt invisible? You might have felt that way, but it doesn't mean it has to stay that way. O'Connor made his own breaks and now he's letting dozens of kids know every year that they do indeed matter. According to his kids, stars and subs matter equally in his program and that explains why so many claim he's truly one of the best there ever was. And did I mention—he wins?*

# VAL ACKERMAN

- ★ SERVES ON THE EXECUTIVE COMMITTEE OF THE NAISMITH MEMORIAL BASKETBALL HALL OF FAME
- ★ FIRST PRESIDENT OF THE WNBA
- ★ 4-YEAR STARTING BASKETBALL PLAYER FOR THE UNIVERSITY OF VIRGINIA

If God had intended man to engage in strenuous sports, He would have given us better knees.

—DR. ROBERT RAY,
*orthopedic surgeon*

My dad was a college and then a high school athletic director, so naturally he was all for my brother and me getting involved in sports. I grew up in Pennington, New Jersey, and we always had a game going on around my block, which was a blessing because, as a girl growing up in the 1960s, there were no organized sports for girls until high school.

## CUTWORTHY

Without a sport to play and being too scared to play with the boys, I decided to go out for cheerleading in junior high. But I was cut, yes, cut, from my first organized team, and it was devastating. Here's the twist: I deserved to be cut. I was not the cheerleader type. Other girls were better and perkier. I just didn't have it. But being cut did teach me that when you do something you love, it shows, and if you don't have a passion for it, that shows, too.

## AND SO IT BEGINS

In high school, I had athletic opportunities and I took them. I played varsity basketball as a freshman, field hockey, ran track, and in the summer I swam. Basketball was my best sport. My dad was always my biggest supporter, and I still have a very vivid memory of shooting one thousand shots a day with him. As a sophomore, the accolades started to come. I was chosen All-County and had a few really big games. I would end up a one thousand-point scorer, and I knew I wasn't done playing. Back then, there really wasn't any college recruiting for women's basketball, but I knew I wanted to play hoops at the next level. So I walked from college to college with my scrapbook going from Duke to North Carolina to Virginia. Only Virginia offered me money and I knew the coach, so I went there.

## MOMENTS

I had a lot of great moments because I spent countless hours alone, working with the ball, shooting and training. I was not going to the mall or watching TV. I was training. Yeah, I was talented, but without the solemn hours I put in I wouldn't have scored one thousand points in high school and at UVA. No way!

## HOW HOOPS HELPED

First and foremost, I learned to manage my time. I learned to prioritize. I knew when to take my books on the bus and how to study efficiently, which are skills I've used every day since. I also knew about sacrifice, which is why I waited until my thirties to have kids.

## IT'S NOT WILL YOU FAIL, BUT WHEN

Talk to any successful person and they'll tell you they have failed. How you deal with failure and what you learn from it will determine how far you'll go.

*So, Val, do you have a story to share?*

My coach and I were from New Jersey and she decided to set up a homecoming game of sorts versus Temple in Philadelphia. Well, with my family and friends in the stands and the game down to the wire, with us up by one point, my coach asked me to inbound the ball with just seconds left on the clock. How could I screw that up, right? I was supposed to find a guard to run out the clock and then we'd win the game. Well, I inbounded the pass, it got picked off, Temple scored and won the game. I was inconsolable. I went from the ultimate high to the ultimate low. I went home for Christmas break and my dad said, "The sun will still come up tomorrow. These things happen. You have to shake it off." Well, the sun did come up and I was able to shake it off, and I never threw away another inbound pass the rest of my career. Our season really picked up the second half and we had a fine year.

## WRAPPING IT UP

All these things helped me have a successful career and work my way back into sports, leading the WNBA to a steady, strong start. I was able to do well because I was used to the scrutiny, used to taking chances, and I enjoyed the challenge, just like the days when I was out there sweating on the court.

**MY WRAP**

*She may be one of the most high-profile success stories in women's sports, but Val did not make it without the tribulations. Hopefully, her honesty will inspire others to persevere. And above all, just play.*

# SENATOR BEN NIGHTHORSE CAMPBELL

★ **U.S. SENATOR, 1994–2002**

★ **COLORADO CONGRESSMAN, 1987–93**

★ **CAPTAIN OF U.S. OLYMPIC JUDO TEAM, 1964**

★ **3-TIME U.S. NATIONAL JUDO CHAMPION**

The rigid volunteer rules of right and wrong in sports are second only to religious faith in moral training.

—PRESIDENT HERBERT HOOVER

I was a troubled kid. I was born into a dysfunctional home and was placed in an orphanage when I was young. I was a high school dropout. I did the kinds of things you wouldn't want your youngster to do, and if I hadn't gotten involved in sports I would have been serving in a different kind of institution than the U.S. Senate.

I found judo in 1949 and it saved my life. I got into judo totally by chance. I was working in a fresh fruit packing plant and I met a group of Japanese kids and we bonded. It was right after World War II and they were experiencing some discrimination, as you would imagine. I knew the pain of that, being a mixed-race American Indian myself. One day they asked me to come down and learn some judo. At the time I was fourteen and I wanted to learn just so I could roughhouse and throw someone on their tail. As I grew in the sport, the support I received turned into winning a national championship and later a spot on the U.S. Olympic team.

Judo gave me a healthy lifestyle I've maintained, since I still don't smoke or drink. And I still train three to four times a week and that's a discipline that's transferable to other facets of my life.

### JUDO: LEARNING TO LOSE

In judo, if you win a match, a fresh opponent comes out. No matter how good you are, when fatigue sets in, you will lose. The Japanese think that you learn as much by losing as by winning. On weekend tournaments you might fight twenty-five or thirty matches in two days. I averaged about one thousand matches a year. I won so many matches that I stopped keeping the trophies. Many kids I taught judo to went on to complete and healthy lives instead of lives of crime and despair. I remember these kids sleeping in the gym after a workout and then taking them for breakfast and maybe golfing. Linda (my wife) and I made an impact with those kids and it feels great even today

### HOW TO KNOW IF YOU'RE GOING TO WIN

I would pick up on little things that gave away whether or not I had the ability to beat my opponent. I have found that whoever speaks first or whoever walks over to introduce themselves to the other guys is the one who is going to lose. It meant they did not have the frame of mind to win. It's as if they gave up their match before the fight.

### ADAPTING JUDO TO WASHINGTON, D.C.

> The Willow bends in the hurricane while the sturdy Oak is destroyed.
>
> —BEN NIGHTHORSE CAMPBELL

When it comes to lawmaking, I try to never meet force and aggressive debate with force. If I sense confrontation, I relax a little bit and bend with it. The force will go by me and then I will survive. An example of a project that took all my judo restraint was the Ute Indian Water Right Settlement Act. It took sixteen years and just about all my judo discipline to make that happen. Others were opposed to getting this Indian tribe regular water, but eventually I got it passed.

### HOW DO YOU PLAY THE GAME?

You don't win the day of the tournament, you win months if not years before the tournament. I saw many people who were better and stronger than me lose. What they didn't have was the heart to train day after day after day, when no one else was around. They didn't have the drive to leave their buddies in the bar and go run and lift and spar. The training is the important part—not the winning.

### MY WRAP

*It's hard to picture a person doing more with less than Ben Nighthorse Campbell. He's the type of person who had every excuse to be on the wrong side of the law, but he ended up being a world-class athlete and an outstanding congressman and senator. His theory on sizing up his opponents by how they approached him before a match is fascinating. It shows how many different ways an athlete can look to gain an advantage and how often the mind decides if you win or lose.*

*Cheney, second row, right*

# DICK CHENEY

★ **VICE PRESIDENT OF THE UNITED STATES, 2001–PRESENT**

★ **SECRETARY OF DEFENSE IN THE GEORGE H.W. BUSH ADMINISTRATION, 1989–1993**

★ **U.S. CONGRESSMAN FROM THE STATE OF WYOMING, 1979–1989**

★ **WHITE HOUSE CHIEF OF STAFF IN THE GERALD FORD ADMINISTRATION, 1975–1977**

Coaches who can outline plays on a blackboard are a dime a dozen. The ones who win get inside their players and motivate.

—VINCE LOMBARDI,
*legendary Green Bay Packers coach*

My mom was the one who taught me how to compete. I watched her play baseball and softball, which was something my family could do together So, it was only logical that when I was old enough, I think I was nine or ten, I would join Little League, which I just loved.

Once I reached high school, I started playing football. I was captain of the team and played linebacker and halfback my senior year. We were the largest high school in the state and in order to play teams at our level we would travel to places like Scottsbluff, South Dakota, and Grand Junction, Colorado. We played solely for the love of the game and believe me, there was no danger that I was going to get into the NFL.

We had a great coach named Harry Geldien. Not long ago, I ran into him at a team reunion where we all reminisced about those years. The things that stood out for me were in a humorous vein. I was reminded of what one of my coaches once said to me: "Cheney, you are a great mudder. You're great in the mud. The only problem is, it never rains in Wyoming."

Over the years, I guess we began to think we were better than we were. One of the guys found the game films in black and white Super 8, with no sound, and we had a showing at the hotel before the actual reunion. It was depressing.

*Why?*

Because we weren't nearly as good as we thought we were. It was sobering because after all those years our memories had smoothed over the rough edges, and this was the reality brought back to all of us.

Football season was something we looked forward to and to some extent, at least, it was the highlight of the year when we'd embark on that nine-game season. It's where I first learned to give it everything I had, because you couldn't go back and replay the game later. It was hard work, you had to commit to the enterprise, and that's where that concept was drilled into me.

## FRIENDSHIPS

Many of my teammates from those days are still my friends today. One of them, Joey Meyer, with whom I went to high school, was my roommate in college, and even dated my wife—before me, of course. He even stood up for us at our wedding. What I remember best about him is that he was the guard and I was the halfback and I had a particular view of his anatomy before I really got to know him in all the other ways friends do.

## CHENEY THE VICE PRESIDENT ON CHENEY THE HALFBACK

I was not an outstanding runner and I needed all the help I could get. I played it by the book and took pride in my play, but I was realistic about my skill level: I was an adequate high school halfback. Our best year was when I was a junior. We tied for the state championship, and believe me, it was a big deal.

*Time for an audible. Like the president, George W. Bush in my first book, I have to go to a second source to find out who Dick Cheney was as a football player and how it helped him become who he is today. Meet Joe Meyer, his high school friend, teammate, and college roommate, and now Secretary of the State of Wyoming.*

I knew Dick Cheney as a guy who loved to hit, and in fact we loved to hit each other. Dick was tough. He would dig his heels in and I never saw him flinch. I think we were similar in that we both wanted to test ourselves to see what we could do.

One moment we'll never forget is how we made it up to the varsity. We were on the sophomore team, at training camp, lining up against each other. The drill was that one guy got the ball and the other guy was supposed to tackle him. We did it a few times and then the varsity coach was called over to watch us. Both of us just banged each other so hard, so many times, that we got moved up to varsity that day. Dick never quit. He loved to bang. On offense, he might not have been fast, but he could block and, like I said, he loved to hit people. Our senior year, we were 2-6-1 and we both made All-State. Dick can "aw, shucks," all he wants, but he could play. Keep in mind, we didn't weigh more than one hundred fifty pounds at the time. We weren't interested in the glory, we just wanted to see how good we and our team could be.

**FINAL THOUGHTS, MR. VICE PRESIDENT?**

Sports gave me something I could succeed at early in life. They also helped me develop the self-confidence and discipline I like to think I have today.

**MY WRAP**

*Look at it any way you want, but I think there has to be something about the way Dick Cheney played then and the kind of person he is today.*

# DARRELL WALTRIP

- ★ **LEAD FOX SPORTS NASCAR ANALYST**
- ★ **ONLY DRIVER TO WIN $500,000 OR MORE IN A SEASON 18 TIMES**
- ★ **TIED FOR THIRD WITH BOBBY ALLISON ON ALL-TIME NASCAR VICTORY LIST WITH 84 WINSTON CUP (NOW KNOWN AS NEXTEL CUP) VICTORIES**

If the lion didn't bite the tamer every once in a while, it wouldn't be exciting.

—DARRELL WALTRIP

I was a track star and I loved it because it was all about heart. It was the kind of sport that if you put in the time and trained every day, you could do well. I wasn't interested in basketball or football, so track became my thing, and it gave me the confidence to try other things, like go-kart racing. It also gave me an identity because I was not an outgoing person.

*Could this be true? Darrell Waltrip—shy?*

## MOMENT: OPPORTUNITY BLOWN

I broke a Kentucky state record in the 880, but at the state championships, I blew it. I went up there with some friends and as soon as we pulled in we went out on the town. I should have been looking to break the state record again, but instead I went out with my friends. It was embarrassing, not because I lost, but because I didn't give myself a chance to win. I swore that nothing like that would ever happen again in any sport, at any time.

## DETHRONING THE KING

Don't beat yourself.

—DARRELL WALTRIP

I stopped running track after that state meltdown and began racing cars. In 1979, I was doing so well, I just set my aim on the king—Richard Petty. I was winning the point standings and had a one-lap lead on the field. All I had to do was coast to victory, but instead I chose to send a message and tried to lap everyone again. My crew was yelling at me to slow down, collect the check, and go home, but I wouldn't listen. Sure enough, I went into a turn too high, spun out, wrecked the car, and I was out of the race. Everyone was mad at me, and my confidence was shattered. Soon my point lead had dribbled down to two, as Petty just chipped away. By the time the final race came, we were all uptight. In the end, I lost the title by eleven points. I beat myself and it was hard to live with. That 1979 season, which taught me to be humble, left an indelible mark on my heart and soul.

## JUNIOR COMES UP BIG: THE DIFFERENCE ONE MAN CAN MAKE

I had a humiliating off-season, but fortunately the legendary Junior Johnson called me to race his car the next season. That brought my confidence back. I thought, "If Junior Johnson still sees me as talented, then maybe I can bounce back from this." I went on to win it all in 1981 and 1982, which turned out to be one of the best back-to-back seasons in NASCAR history. All because I learned from 1979, combined with Junior's staunch support for me and a belief in my future.

### "I WAS A NOBODY"

Throughout high school, I was invisible and just wanted to be liked . . . and I wasn't. I still have that feeling today. I am still a people-pleaser. I care what people think. This made my early years in racing sooo tough because I was booed louder then Kurt Busch and Jeff Gordon combined. What I needed to do was find the Lord and stop thinking about me and instead look out for other people. And when I did, my life immediately changed course.

### AS A BROADCASTER

I try to approach the broadcast booth like I would a team. We all think the same way in the Fox booth. I know and they know instinctively what my strengths are and we all just try to make the other guy look good. It works. Why are we like that in this cutthroat business? Because we all come from team sports and we know what works and what doesn't.

### MY WRAP

*Notice the difference one person can make in someone's life. In this case it was Junior Johnson. You might have the same opportunity with an employee, a coworker, a mentor, or one of your kids. Darrell's life has been on hyperspeed because of his racing, but I'm sure on some level most of us can relate.*

1948

# HARVEY MACKAY

- ★ MOTIVATIONAL BUSINESS SPEAKER AND DIRECTOR OF ROBERT REDFORD'S SUNDANCE INSTITUTE
- ★ BESTSELLING AUTHOR OF *SWIM WITH THE SHARKS WITHOUT BEING EATEN ALIVE* AND *BEWARE THE NAKED MAN WHO OFFERS YOU HIS SHIRT* (SOLD 10 MILLION COPIES WORLDWIDE)
- ★ FOUNDER AND CHAIRMAN OF MACKAY ENVELOPE COMPANY

Baseball gives you every chance to be great. Then it puts every pressure on you to prove that you haven't got what it takes. It never takes away the chance, and it never eases up on the pressure.

—JOE GARAGIOLA,
*baseball player and Hall of Fame broadcaster*

I was fourteen years old in 1946 and the local minor league baseball team, the St. Paul Saints, had a contest each year to see who could sell the most tickets. The prize would either be to hit the first ball or throw out the first pitch. I had an idea. What if I went around and told the businesses to buy tickets from me so I could send orphans to a Saints game? Who could turn an offer like that down from a fourteen-year-old kid? I put hundreds and hundreds of orphans in that stadium and won the contest. I chose to swing at the first pitch in front of twenty-one thousand people, while also getting my photo on the front page of the paper. I still remember all the businesses that bought tickets from me. It's how I first learned I could sell and be a leader, and the first time I mixed sports and sales. It certainly wasn't the last.

I was always a sports junkie. For three years in a row, I was the St. Paul City Gold bowling champion. I three-putted the 72nd hole to lose the Minnesota State golf championship, and then I went on to play at the University of Minnesota. I was captain of my baseball and basketball teams. I was also state junior ping-pong champion, so I guess you can say sports dominated my young life.

### THE MOMENT: "THERE ARE NO GIMMES"

When I was fifteen, I weighed all of one hundred thirty-five pounds and I found myself playing in the St. Paul Open golf tournament, one of the top tournaments in the country at the time. The best of the best, like Sam Snead and Ben Hogan, played in that tournament. I got in because my handicap was 2 and that's what you needed to qualify.

I was having an incredible experience and was playing well when I got to the fifteenth hole. I looked up and saw twenty thousand people, including my family, standing there, watching me. It was a par five. I had a 2 iron in my hand and I rimmed the cup for what could've been a double eagle. I couldn't even see the ball, I just saw the crowd's reaction, and I knew that something special had happened. When I got up there, I saw that the ball was just ten inches from the cup. I couldn't believe it. With all those people watching me, I didn't want it to seem like I had to think through a shot like that, so I just went for the tap-in and I missed it. I missed a ten-inch putt in front of all the cameras, all my family, and the twenty thousand other fans. Ouch!

I still think about that today. But it taught me a lesson that I still adhere to today: never to take anything for granted. Another thing I learned is to get over worrying about how other people think I look. I should have gone into my stance and sunk it for a 74. Instead, I shot a 75 and missed the cut. Another lesson: always pour it on, whether you are winning or losing.

Years later, I lost a union vote that I worked so hard to win. I didn't care because I knew I poured it on to win that election, and just because it didn't happen didn't mean I failed. This was the real payoff from one bad putt!

## FOCUS

A second indelible moment for me happened once again at the St. Paul Open, only this time I was in college. The day before the tournament, the future legend Gary Player came over to my campus to take a lesson from my coach and I met him. What a thrill it was to meet a man that talented up close. What was even better is I got a chance to go out with him that evening and we had a great time. The next day, I see Player about to tee off on the first hole with almost no one watching, other then the scorekeeper. I dropped my club and ran up to him and said, "Gary how are you? Wow, did we have a good time last night." He stared, steely-eyed, straight ahead. Without breaking stride he said, "Harvey, please do not talk to me. I must concentrate. I will see you when I finish." I left, tail between my legs, and didn't see him again for twenty-five years, during which time he becomes one of the greatest golfers ever. One day, I happened to see him in a restaurant. I reintroduced myself and explained how bad I felt all those years ago. His wife interjected, saying, "Don't feel bad. He won't even say hello to me when he sees me on the course. It's just what he needs to do." So, Player taught me focus, and the art of concentration!

## MY WRAP

*Most of us have the opportunity to pick up the lessons Harvey did, but we don't always have our antennas up to receive them. For Mr. Mackay, school is always in session despite his amazing business, speaking, and writing success. Looking at what he did in sports, I have to believe that had he chosen to do so, he could have turned pro. But pro or not, his message is clear: compete hard and play fair, and if the game doesn't go your way, the people you meet on the course or on the field might just provide the win you seek somewhere down the line.*

# JOHN LYNCH

★ WITH WIFE LINDA, FOUNDER OF THE JOHN LYNCH FOUNDATION, TO DEVELOP QUALITY YOUNG LEADERS

★ 6-TIME PRO BOWL SELECTION

★ SUPER BOWL XXXVII CHAMPION, 2003

★ NFL SAFETY, TAMPA BAY BUCCANEERS, DENVER BRONCOS, 1993–PRESENT

No coach ever won a game by what he knows, it's what his players have learned.

—AMOS ALONZO STAGG,
*University of Chicago football coach*

The first example of going the extra mile in sports came when I was ten years old. Before school, my brother and I would go to the gym and work out. It was fun. I knew no one my age was doing it, and most of all it was a bonding time with my dad. It was his way of teaching me the work ethic to be successful not just in sports but in life.

Oddly enough, my first sport wasn't football or baseball, it was swimming, a great sport in which to learn a good work ethic. At five years old, I was swimming at six-thirty in the morning, which I kept up until I was nine. I knew, even then, that I loved competing, because I wanted to make my parents proud. They were not that interested in my success, just how hard I worked. (By the way, dad says that at the age of six, I set a California state record for the fifty-yard freestyle for eight-year olds.) Now, as a parent myself, I understand that there is nothing better than seeing that kind of drive in my own kids.

After swimming, it was on to soccer and football. I remember making travel teams in both sports, swapping out uniforms in the car going from one game to another. It was crazy, but I just loved it!

**IMPACT COACH**

If you want to talk about an impact coach, believe it or not it wasn't in football or soccer, but in basketball, courtesy of the Boys and Girls Club. His name was Ron L.—I never did learn his last name—and he was the basketball coach. He knew I wasn't the best basketball player, but he was the first one to say that I had something special. "You know how to compete," he said. To be singled out at such a young age meant a lot to me, so much so that I still remember it today.

**DREAMS**

I am so lucky because my parents encouraged me to dream. So many parents today want to temper their kids' hopes in order to avoid disappointment, but my parents made anything seem possible. It also helped that my dad made it to the NFL for a brief stay with the Steelers. In my mind, he was a part of those great Steeler teams, and I just wanted to be like my dad and that meant going pro.

With every dream, my mom and dad emphasized hard work—that was the route to success. Do the extra things, the little things, and you can make it happen. That was their lesson. Even today, that philosophy pays off. I'm in my fourteenth year in the NFL and every day, every practice, is still a challenge.

*Just ask his teammates with the Bucs and the Broncos and I'm sure they'll agree. I sure do.*

Every day, teammates are trying to take your job away, or the opposing team is trying to expose you. That's the challenge that gets me going every day. To this day, I just love going to work.

### LIFE HAS NOT ALWAYS BEEN DREAMY

Going into my senior year at Torrey Pines High School, I was having great academic and athletic success. After all, what could be better than being the quarterback of a highly regarded football powerhouse? So my coach decided to raise the stakes. He recruited a coaching legend—who'd worked with John Elway as a high schooler and was one of the inventors of the run-and-shoot offense—to change our offense. For a quarterback, this was an incredible opportunity. In fact, there were scrimmages where I threw for five hundred yards. Recruiters were calling nonstop. Then, it all stopped. In the first game of the year I broke my ankle. The college calls, the attention by the press, all came to a screeching halt.

It was only then that what my parents had preached to me from day one about grades coming first and sports second really hit home. After the injury, I thought the only way I would get into college was through academics. Thankfully, I already had good grades, because if I didn't it would have been too late. Then I got a break. Notre Dame offered me a scholarship and then some other schools followed their lead. Suddenly I had options. I chose Stanford. You'd think I'd have been home free, right?

Wrong!

### BIG DECISION, BIG RISK, BIG REWARD

At Stanford, I was slotted as the number-two quarterback, but I just didn't move up. What didn't help is that I played baseball, too, so I wasn't able to play spring football. My junior year, I thought for sure I would get the starting job. Dennis Green (the former Minnesota Vikings coach and now Arizona Cardinals coach) just handed it to another player and I was devastated. I thought I had to transfer or make the decision to just play baseball.

## TIME FOR PLAN C

I went to see Coach Green and said, "Just put me on the field." At first, he resisted, but eventually he gave in. I left his office with him agreeing that I'd try the safety position. It was a tough transition. I made some mistakes, as I mostly played nickel back that year. I was set to try to make it as a pro baseball player—the Marlins drafted me in the second round.

*Side note: he was the second player and first pitcher drafted in Marlins history.*

Then something unforeseen happened. Coach Green went to the Vikings and Bill Walsh came out of retirement to take over the team. He watched my tapes, called me in, and said, "Look, I know you could play major league baseball, but I think you could be a great safety for us and in the NFL."

He took the time to edit a tape of the four games I started and spliced in plays from Ronnie Lott and showed me the similarities. Well, he sold me. I told the Marlins I was staying and started for Stanford my senior year. I got drafted by the Tampa Bay Buccaneers and fifteen years, six Pro Bowls, and one Super Bowl later, I'm still playing.

## SO, WHAT DID HE LEARN?

To grow up. My dad always wanted me to be a quarterback. He thought I could be a great quarterback and wanted me to transfer out of Stanford when I wasn't chosen to be the starter. I thought I knew what was best, and it turned out to be the right decision. At one point we all have to figure out who we are and why we're playing. That was the point at which I started playing for me and not for anyone else. There's an excellent chance I would have made it as a quarterback, but I'd had it with waiting around and I had to follow my heart. In the end, it worked out for me.

*I'd say so, and if you ask me it's likely he'll be even better as a broadcaster in his next career. Just like his Boys and Girls Club coach who singled him out and got him to believe as a teen, Walsh sold him on his own dream, just before he audibled to baseball, broadcasting, or something else.*

## HAPPILY EVER AFTER? NOT QUITE

After twelve great years with Tampa and a Super Bowl win over the Raiders, I received another wakeup call. It didn't come as a complete surprise because you know you're in trouble when the people close to you stop talking to you, and that's what happened with the Bucs. They thought I was finished as a player and didn't want to pay me what was left on my contract, so they let me go. The same thing happened to Joe Montana, Emmit Smith, and Junior Seau, so why should I have been surprised? Yet I was. I took it very personally and was devastated for a few days, but when interest started coming in from other teams around the league, I felt a lot better. Thankfully, it's worked out great with the Broncos. Returning to the Pro Bowl was special for me.

## HIS PROUDEST MOMENT CAME FROM A CRUSHING DEFEAT

In the 2005 season, we were one step away from the Super Bowl. All we had to do was beat Pittsburgh, who had been on the road five straight weeks. We were feeling good about our chances after knocking out the defending champion Patriots, which set things up for a title game at home. We lost, and I am just not a good loser, as hard as I try. This one might have been my last chance at the Super Bowl. My family was outside waiting for me along with a few thousand fans hoping for an autograph. I wanted to just grab my friends and family and duck out the back door, and I decided to do just that when halfway through the back tunnel, while holding my son's hand, I just stopped and said to myself, "What kind of example is this for my son?" I did an about-face and headed right for the Steelers locker room, and as hard as it was watching the champagne flow and the celebration in full swing, I worked the room and shook the hands of Jerome Bettis, Bill Cowher, and Hines Ward and told them to go win the Super Bowl. I felt good about that, but even better that my son saw it. I can tell my son how important it is to be a gracious loser, but this was a way to show him. Train hard, play to win, but if you lose, do it with class. He's just seven, but I hope it shapes his life forever, because in the end, that moment shaped my life.

The concept of work hard off the field to be successful on the field as an effort to develop young leaders is the essence of my foundation (johnlynchfoundation.org), founded by my wife, Linda, and me. She is an outstanding athlete who played tennis at USC and went on to compete on the satellite tour for a few years. We both felt the lessons we earned as student athletes were so valuable that not only did they help me make it as a pro, but they're the same lessons we use to raise our kids. Sports are such a tangible way to learn life skills. I can tell people all day when things don't go your way and you get knocked down that you have to get back up, but without a real-life experience for them to draw from, the concept is just a bunch of empty words. When you're a defensive back and the receiver you should be covering catches a ball over you and goes in for a touchdown and your team is trailing because of you, your mettle gets tested. If you stay down, if you let your confidence get shaken, well, then, you're in real trouble because five minutes later you'll have that same guy to shut down, and if your head's not right, you're done and so is your team. They say you have to be quick to forget, and I've learned to be quick to forget in anything I do, because what I do or don't do against the Chiefs is not going to help me against the Raiders.

## MY WRAP

*John is not only an athlete, but he's a special person who's had it anything but easy. From his broken ankle in high school to riding the bench in college to getting cut as a pro, he had moments that would rattle all of us (including him) to the core. I'm sure many readers have at one time been fired, cut, let go, rejected, or turned away. There is glory in just fighting through it, even if the results don't meet your goals. And as you learned from John's last story after the AFC title game, the learning doesn't stop just because we've supposedly grown up.*

# CHRIS EVERT

★ 18 GRAND SLAM SINGLES TITLES—AUSTRALIAN OPEN, 1982, 1984;
FRENCH OPEN, 1974–75, 1979–80, 1983, 1985–86,—A RECORD
7 TIMES; WIMBLEDON, 1974, 1976, 1981; U.S. OPEN, 1975–78, 1980, 1982

★ SINGLES WIN-LOSS RECORD OF 1,309–146 (.900) IS THE BEST OF
ANY PROFESSIONAL PLAYER IN TENNIS HISTORY

★ 4-TIME ASSOCIATED PRESS FEMALE ATHLETE OF THE YEAR

★ PROFESSIONAL TENNIS PLAYER, 1972–1989

The moment of victory is much too short to live for that and
nothing else.

—MARTINA NAVRATILOVA,
*legendary women's tennis player*

I was really a brat when I first started playing tennis as a kid. I loved the game from day one, but that doesn't mean that I was always the cool, calm, collected person you saw as a pro. I would mishit the ball and then throw my racquet. I'd voice my frustration loudly. You name it, I did it on the practice court.

One day my dad pulled me aside and calmly said, "Don't let your opponent know how you're feeling because they will use it to their advantage. If you hide it, they'll never know how you're feeling and you can use that to your advantage." So I started to mask my emotions, and sure enough, I would notice that my opponents would get emotional and I'd capitalize on it.

*Was that it? Just one conversation with Dad and you got yourself under control?*

There was one other time when it was my brother who gave me a reality check when I was seventeen. I was working out, playing with him, and I had a hissy fit, which I guess was pretty bad. He stopped, looked at me, and said, "Chrissy, just listen to yourself." And then he walked off the court and went home. I couldn't believe it. No one ever walked off the court on me!

But I did what he said: played back in my head my comments, as I sat on the court alone, and I came to the same conclusion. I was being a b____ and he just put me in my place. At that moment I began to realize I had to be sensitive to other people. Up until then, I'd probably been too focused on me.

*You mean, like every seventeen-year-old?*

The fact that I can still picture my brother's face and recall the incident shows what an impact the moment made.

### LET'S GET THIS STRAIGHT

I liked to play tournaments—I loved the pressure—but I didn't like to practice. And yet I did it, and I did it as much as anyone.

*Doesn't this make us feel better?*

I liked the tournaments because I developed a knack for being calm under pressure and I could sense that most of the other players used to tighten up and often played nervous. This always gave me an advantage.

*Wait, you know the secret to conquering pressure?*

The ability to deal with pressure and stay calm came from playing a lot of tournaments when I was young. It was not "now or never" for me. I taught myself to be in the moment and, for the most part, that worked for me.

*So, can you use this quality in real life?*

When I broadcast a tournament or make a speech, I use the same technique. I prepare from the heart and I study. In tennis, I learned the draw. If it's a speech, I prepare myself on the topic. I'm not book smart and I don't know every fact from my seventh-grade history class, but I do know about life experiences, about feelings, and about mastering emotions.

*Can you point to one match in your career that taught you the most?*

There was one match against Tracy Austin in the 1980 semifinals of the U.S. Open that comes to mind. I'd lost five straight times and everyone was saying how she was the new Chris Evert and that I was not supposed to beat her. Par for the course, I dropped the first set 6-4. I was playing well, but I was still losing! I knew I had to take a risk and, just like that, instead of being the counterpuncher I became the aggressor. I changed my style. I came out of my shell and I won the next two sets, 6-1, 6-1, and won the Open. It worked, so I knew it was the right move.

Even when I was down in the match I did not give up and did not stop believing in myself. It was pivotal on both fronts for me and still pays off today. It not only got me the win but prolonged my career ten years because I took a risk. Overall, I hated to lose more than I loved to win, which fundamentally is why I adapted in that match and the many others that followed.

**AND DOES THIS HELP TODAY?**

My brother and I have a tennis academy in Boca Raton that we've poured millions into. For the first five years, building it and making it grow was tough. The economy was not good and we were losing a lot of money and, for a while, we started to question what we were doing. But instead of bailing out, we fought through it. We had the confidence to stay the course and now it's rolling along, doing great. It reminded me of the same way I played the game. On the court, you're really more in control of your destiny, but the feeling and focus inside are the same.

## MY WRAP

*I don't know of a more dignified, intense, or in-control player in any sport than Chris Evert, and she's still like that today. It's fascinating to find out that she had to grow into her unflappable self and that she is, indeed, human after all. Keep in mind that her dad was a pro and her sibling played, too, but it was she who decided to pursue the game to the ultimate pinnacle. And as much as she loved to play, she hated to practice, which I know many of us can identify with. If you're a player or you have a child who's a player and you want to get into emotional shape, to control your rage, try telling yourself Chris's story. I don't think you'll find a better role model.*

# SUGAR RAY LEONARD

★ **WORLD BOXING CHAMPION AS A WELTERWEIGHT, JUNIOR MIDDLEWEIGHT, SUPER MIDDLEWEIGHT, LIGHT HEAVYWEIGHT**

★ **GOLD MEDAL WINNER, 1976 OLYMPICS**

★ **GOLD MEDAL WINNER, 1975 PAN-AMERICAN GAMES**

★ **THREE-TIME GOLDEN GLOVES CHAMPION, 1972–74**

Float like a butterfly, sting like a bee.

—DREW "BUNDINI" BROWN,

*cornerman and assistant trainer to Muhammad Ali*

If you're looking for my sports moment that still pays dividends in my life today, it has to be the first fight I had with Tommy Hearns back in 1981. It was the toughest fight of my life. It was over a hundred degrees in the ring and Tommy was getting the best of me. To be able to put it all together and knock him out in the late rounds reassured me that I could do whatever it takes to win.

*Wait a minute! Sugar Ray was a 1976 Olympic gold medal welterweight champion and he still didn't think he was truly a champion until five years later? What's wrong with this picture?*

Having the ability to overcome obstacles in the business world and as a husband and a father, you learn to maintain your composure. When times are tough, I'll imagine those moments in the ring to bring a balance to my life that helps keep any stressful situation in perspective.

### HOW DOES SUGAR RAY PLAY THE GAME?

I boxed aggressively, fair, and smart. I never tried to fight to the level of my opponent. I tried to dominate, to annihilate my opponent, even my own brother. It was all instinctive and I'm sorry to say it, but I don't think you can instill that mind-set. It doesn't mean that I was always successful, but I tried. For example, I've had the same mind-set in golf, and yet I haven't had close to the same success I had in the ring. But I still go after it the same way, whenever I can.

### AND WHERE DOES THIS MIND-SET COME FROM?

There's no doubt about it, I got this drive from my mom. The best example I can give is that one day my mom was driving the whole family to the store uhen the car went flying violently off a sharp curve in the road. My sister ended up in the front seat. All the doors were crushed closed. I was semiconscious. I remember looking up into the driver's seat and seeing the steering wheel bent up and my mom gone. Suddenly my door flew open and there was my mom, her lip hanging off, grabbing us, pulling us out of the car, and then walking us down a dark road to our neighbor's house. Nothing, but nothing, was going to stop her from saving her kids. And nothing was going to stop me in the ring or, for that matter, in anything I do. I imagine that day and my mom almost every day of my life. It keeps me going.

## THE MESSAGE?

Try. Try to do things you are not sure you will be successful at. Even if you fail, you grow. Losing makes better winners. It's all about the process. I loved my career, but I am by far happier right now than I ever was during my years in the ring. I'm coming up on fifty years old and I am finally using all those lessons from the ring in my life. I'm giving back to others, so they don't make the same mistakes I made. And hopefully, they'll get to feel some of my success in whatever they do.

## MY WRAP

*I've had the opportunity to interview Ray many times over the past fifteen years, and through it all I do not remember ever seeing him so happy and content. Nothing is better for an interviewer than to talk with a guy who enjoys looking back on his life and his career and then bringing you inside. He's an example of a person who never stops learning. Ray seems to get the same thrill inspiring others as he did winning those championship belts. He wants to win, and while he knows that we all must lose sometime, that doesn't mean we should let anything stop us from trying.*

# HANK WILLIAMS JR.

★ WROTE AND PERFORMED OPENING THEME OF *MONDAY NIGHT FOOTBALL*

★ GRAMMY AWARD, 1989

★ CMA ENTERTAINER OF THE YEAR, 1987, 1988

★ COUNTRY/SOUTHERN ROCK SINGER AND SONGWRITER

I don't want to be a hero; I don't want to be a star. It just works out that way.

—REGGIE JACKSON,
*Hall of Fame baseball player*

I was one of those kids who grew early, which was great playing football because I could just slaughter everybody. For a while there, I was unblockable. One day, with my mom watching from the sidelines, the governor's son came out of a huge pile-up with his wrist bone sticking out of his skin. My mom had a full view of this and she grabbed me after the game and said, "That's the end of this." So it seemed like I was done with football at fifteen. Of course, I still played, but it freaked out everybody's parents.

## READY TO GO

I approached music with the same tenacity I did playing football. I think guys like Kenny Stabler and Ben Rothlisberger are the same as most musician in terms of mind-set. We all have the endurance and know we have to play in all types of conditions, and play through fatigue. I may say that I don't want to play in Pierre, South Dakota, tonight, or Ben could say, "I don't feel like playing in Green Bay today," but we both do, and we do it the best we can. It's all mental.

What I've learned over the years is that endurance is the key. You have to have endurance to be able to do shows for forty years and still get pumped up enough each time to be able to pull it off. It's something you have to do in sports, too. Sometimes I can't believe these baseball players are going at it for 162 games a year, and that's without playoffs, being judged and cheered and booed on a daily basis. That's a grind.

## THE FACE OF *MONDAY NIGHT FOOTBALL*

*Monday Night Football* was a one-year deal that turned into an American signature. It's allowed me to truly be a part of the sports community. I've come to realize how close music and sports really are. I also had a chance to become good friends with the late Derrick Thomas of the Kansas City Chiefs. He'd go on the road with me, and that's when I really saw how similar our lives were.

## HOW DOES HANK PLAY THE GAME?

You play the way you feel it from your heart. You know what's right and what's wrong, and if you're not comfortable with yourself you are going to lose. Forget the words "I can't." What I want to hear is, "I'm going to." When I get to a town I always try to find out if the local minor league team is playing, and if they are, I like to see if I can take a few swings in batting practice. Like ballplayers, I know how to deal with pressure. I was born Hank Williams Jr., which was like being born Frank Sinatra Jr. Everyone expects perfection. From the day I was born, everyone wanted me to be like my dad. That pressure is a lot to endure. I only started doing well when I did what I wanted, playing music *I* liked and writing the music *I* like best.

## MY WRAP

*All during our interview, I had to keep reminding myself that I wasn't talking to a professional player. Because of his no-excuses mentality and his cadre of friends (mostly pro stars, by the way), I really thought I was talking to a former NFL star. Why? Because he has the elite athlete's mentality of doing whatever it takes to win.*

*Mitchell is wearing number 21.*

# GEORGE MITCHELL

★ **U.S. SENATOR, 1982–95**

★ **MAJORITY LEADER OF THE SENATE, 1989–95**

★ **AUTHORED PEACE PLANS FOR THE MIDDLE EAST AND IRELAND**

Games lubricate the body and the mind.

—BENJAMIN FRANKLIN

M y three older brothers were excellent basketball players. I was not as good as them. In fact, I was not as good as anybody else's brothers. I was known as the Mitchell brother who wasn't any good at sports. Having said that about my ability, I still played basketball through high school and in college. I didn't quit the team or the sport because I was not a star. It was more important simply to remain part of the team.

## A MOMENT I CAN'T WAIT TO FORGET

Although I knew I was not a very good player, I still had pride. My senior year of high school I made the varsity basketball team, although I was not a starter. When we made it into the tournament, my coach called me into his office and informed me that we could only take ten players and that I was not selected for the tournament team. I was crushed. I can remember the moment like it was ten minutes ago, because I felt like such a failure. Now I see it as an important moment because it let me know that there are going to be setbacks in life and that you have to deal with them and move on to other things. That night I went home and informed my parents that I had not made the playoff roster. My father said, "Study hard, focus on academics, and I guarantee your brothers will be looking up at you someday." This was a failure for me, but it was an important failure because it was the first time in my life I was able to get over the feeling of inferiority I had in relation to my brothers, especially in sports. Of course, that feeling did not subside right then and there, but soon after I was able to turn the page in my relationship with them.

## IN PERSPECTIVE

Being cut from my high school playoff roster did not stop me from playing in college, but it did make me realize that sports is not everything in life, just a part of life. Don't get me wrong, I always practiced, and I had some good games, but I had never had a great deal of success.

## HOW IT HELPED

> Although he's regularly asked to do so, God does not take sides
> in American politics.
>
> —GEORGE J. MITCHELL

I don't know of any athlete at any level who wins all the time. Sure, we all
want to win, but in the end it's the competition that matters. It's all about
learning to deal with setbacks and an awareness that teamwork matters.
Sports are just building blocks to a successful life. I used all those personal
lessons in perseverance to broker those international treaties and shep-
herd so many bills through the Senate. I have no doubt that my days on
the field and on the court helped me achieve whatever success is credited
to me during my years in Washington. I learned to shake off setbacks.
When I speak to groups now I say, "You cannot let every loss paralyze
you. You have to work your way through them because anyone engaged
in life will have defeats." The thing I hate to see, and have little patience
for, is someone who can't get over a mistake. Perhaps you can learn about
life through a good mentor or from your parents, but I don't know any
other way for you to experience the real lessons of life accept in sports.

## MY WRAP

*Wow, can I relate to Senator Mitchell's sports saga. And a special note to those
nonstars across the country, working hard as walk-on and bench guys: Senator Mitchell did not quit and was thankful for his struggle soon after hanging
up the sneakers.*

# POPE JOHN PAUL II (KAROL WOJTYLA)

★ POPE, 1978–2005

★ CARDINAL, 1967–1978

★ ARCHBISHOP OF KRAKOW, 1963–1967

More than being concerned with who's going to win the Super Bowl, I feel the Lord is probably more concerned that they might find a day other than Sunday to play it on.

—BILLY GRAHAM,
*Christian evangelist*

AS TOLD TO ME BY HIS LIFELONG FRIEND, JERZY KLUGER:

*"It's pope!! Lolek is the pope!"*

That was Jerzy Kluger's reaction when he found out his childhood friend Karol Wojtyla was named pope back in 1978. Later, when he was summoned to visit his buddy, the headlines in the Rome newspaper read:

POPE GRANTS FIRST AUDIENCE TO HEBREW FRIEND

I first met Karol when I was six years old. I think the first formal soccer match we played was when we were fourteen. We were quite serious when we played, and the games were always tough. I played inside left and he played goalie.

## WHY THE POPE PLAYED GOALIE

The pope played goalie because he wasn't a fast runner. I'm not saying he was slow, it's just that others were quicker.

*Sounds to me like Jerzy's a master diplomat.*

The pope was a great athlete, very strong and, as a goalie, courageous. We all played in normal shoes, not cleats, not sneakers, but walking-around shoes. I remember he would dive facefirst, right at the feet of any player, just to get the ball. He would get kicked, but he never seemed to get hurt and he never came out of the game.

We would practice on the street and we always ended up by the church, kicking the ball against the wall. More than one black-robed priest came out to tell us to go play somewhere else.

*Can you imagine telling the pope to leave the church?*

## CHOOSING UP

Soccer was not an official sport at our school because of the passions it would stir up on both teams. Yet no one interfered when we played after school, as we piled up our jackets and made goals and then proceeded to play a game. When we were making up teams, we always did it the same way: Catholics against Jews. Usually we didn't have enough Jews and it was Lolek (the pope) who would always volunteer to play with us. In Poland at that time, prior to World War II, it was not easy being a Jew. Often, when games were tight, and almost all of them were, you would hear, "Crack the Jew's skull." We learned that the best response was to score another goal. It also gave me great satisfaction to look back and see Lolek giving up his body to keep the ball out of our goal There it was, a Catholic kid putting it all on the line for his Jewish teammates.

## LIFELONG PASSION

Pope John Paul II was also an honorary member of several soccer clubs, including Barcelona and Schalke 04, and he frequently hosted soccer teams at the Vatican.

In *The Simplest Game* by Paul Gardner, Coach Bora Milutinovic described the Mexican national team's audience with John Paul II in 1984 before a friendly match against Italy:

"He gave us a blessing. He held up one hand. Five fingers. The next day the Italians scored five goals against us. We were just happy he didn't bless us with both hands."

## MY WRAP

*Isn't it amazing how often the way one approaches sports indicates who one is as a person? Defined as a goalie when he was young, Pope John Paul II was also described as a kind, generous, courageous, caring person. He didn't end up in the Soccer Hall of Fame, but he will end up in the Human Hall of Fame, better known as sainthood.*

# TONY SIRAGUSA

★ FOUNDER, TONY SIRAGUSA FOUNDATION, HELPING UNDERPRIVILEGED CHILDREN

★ FOX SPORTS NFL BROADCASTER

★ SUPER BOWL XXXV WINNER, 2001

★ NFL DEFENSIVE TACKLE, INDIANAPOLIS COLTS, 1990–1996; BALTIMORE RAVENS, 1997–2001

★ NEW JERSEY STATE WRESTLING CHAMPION, 1985, WITH A 97–1 CAREER RECORD

This is a game for madmen.

—VINCE LOMBARDI,
*legendary Green Bay Packers football coach*

## I'VE GOT TO GIVE MYSELF SOME CREDIT

My first defining moment came on the mat, winning the New Jersey state wrestling championship. I knew I was good, but I was the type of guy who would goof off while others were killing themselves. I always knew what I had to do to be ready, and I had been wrestling since I was five, so I figured I knew what I had to do to win and I guess I was right. To this day, I meet people on planes, in gyms, and in restaurants and they say they remember that I was state wrestling champion. It was big for me, because it helped me as a football player. It was the first time I gave myself any credit. It was also the first time I really focused on anything and applied myself. I was from a small high school, the gym in Princeton was just packed, and to win it was amazing.

I built off the wrestling win and took that winning attitude onto the football field at the University of Pittsburgh. But I blew out both my knees and had to battle right through that.

## HOW IT CHANGED ME

It's always been a challenge for me to prove people wrong. They see me and say, "What can this guy do?" and, in fact, I could do a lot. I like to have fun, but I've proved I can turn it on and off. When I left the field, it was my time, and when I went to practice I had to have fun, just to keep the juices flowing. I would rip on a guy just to loosen everybody up. It was also a mechanism to get people to underestimate me. So many guys live for the game, and when they get out they really can't do anything else and so they struggle. I knew there was more to life than football. My coaches might have been on me to be more serious, but even the guys who were uptight tended to gravitate to me. The people who knew me knew I had the balance down. I knew when to be serious and I knew when to joke around.

## I HAD FUN

Sports, even if it's your job, can be fun. People say work should be serious. Well, the way I see it, you spend the majority of your life at work—why can't you have some fun there? That's what I believe and it's how I approach my job at Fox today!

## WHY I HAVE TO HAVE FUN

My dad passed away at the age of forty-eight and he always talked about what he wanted to do with his life. He wanted to travel, especially to Italy, but he died having struggled and worked his whole life. When he died, I realized how short life can be, which is what a lot of people realized after 9/11. Now I try not to worry too much about stuff. I think that if you're a good person, things will come to you. Smile!

My dad never knew I made it to the NFL, but when I played I did it thinking he was watching me from heaven and I just couldn't let him down. I knew what motivated me and I used that to push it myself. I have the same pressures everyone else has, but I try not to take them so seriously. As a sideline reporter for Fox, I try to make people laugh when the laugh is there, in the same way I tried to give people enjoyment by the way I played when I was out there on the football field.

## MY WRAP

*Tony has one of the great personalities in sports, and he's the first in this book to talk about fun over focus. I think he's as focused as anyone else, but he just doesn't want to let anyone know how much he cared. If you're the cutup at practice, don't think you're just acting like the "Goose," because he wasn't a cutup, he was just breaking up the stress. Tony knew how to straddle the line, and that's why he lasted for twelve years as a player in the NFL. I also owe Tony big-time for the interview he gave me on the field after his Super Bowl win. They cordoned off all the reporters after the Ravens crushed the Giants. He spotted me, told the security guards to let me through, and gave me an exclusive one-on-one. He remembered I wanted to talk with him before he was a national name. Thanks, Tony, and believe me, I'm not the only one who'll never forget you.*

# MARY LOU RETTON

★ **HOST OF TV'S** *MARY LOU'S FLIP FLOP SHOP*

★ **OLYMPIC 5-MEDAL WINNER IN GYMNASTICS, 1984: 1 GOLD, 2 SILVER, AND 2 BRONZE**

World records are like shirts. Anyone can have one if he works for it.

—FILBERT BAYI,
*Tanzanian runner*

There weren't a lot of opportunities for girls in sports in West Virginia, so my mom put us into dance. My sister and I took tap, ballet, and jazz. I knew right away, even at age four, that ballet was not for me, but there was a class called acrobatics and that really turned me on. Later, I ran track, swam, and was a cheerleader, but by age twelve I had stopped all sports other than gymnastics.

Luckily for me, I was born into an athletic, competitive family. After being told I was too little, or that I couldn't do something, or that my time was coming, I just wanted it even more. Instead of feeling low or let down, I felt like I had to show people that I was big enough and good enough to do whatever it was they told me I couldn't do.

## MY MOMENT

Bela Karolyi (*the legendary gymnastics trainer*) saw me at a competition in Reno, Nevada, and pretty much recruited me to train with him in Houston. He said, "Come with me. I make you Olympic champion," and so at fourteen I packed up and left West Virginia for Houston. West Virginia did not have a lot of elite gymnasts—I was essentially it—so I knew I had to step up my training if I was going to reach my full potential. But I was not ready for what I saw in Houston. I was not the big fish anymore. I looked around and saw the talent and I got slapped down right away. Bela only worked with four or five gymnasts. I was put into his group right away and I was at the dead bottom of the totem pole. I had to right my way up, to earn his respect, and I just loved it.

*Most knuckle under that kind of pressure, but Mary Lou thrived on it.*

It was what I needed: to be pushed and challenged, and it's really when and where I started to grow.

## MONKEY MOMENT: "IF I CAN'T, TEACH ME"

I was small and powerful, but the balance beam, which is so different from other events, requires grace, flexibility, and agility. Needless to say, I was not in love with this event, nor did I shine in it. But I needed to. I was known for falling off. When I wasn't falling, well, let's just say I wasn't pretty to watch. It just killed me. My first day at Bela's I saw his wife, Marta, and I remember her saying, "Oh, Bela, why did you bring this little monkey to me?" I would call her Hawkeye, because she always had her eye on me. But I was very coachable and she and Bela respected me because of my work ethic.

## SHE HAD DOUBTS

Don't get me wrong, at times I doubted myself. I was living with a strange family, by myself, in a new school and working out eight hours a day after being used to training just three hours back home. Factor in not being the best anymore and it's amazing I didn't just give up. But after four months, I started getting better and the voice in my head was saying, "I can do this." That's when the parents and the other kids made it really clear that they wanted me to go home, because I began to beat them and get more of Bela's attention, which created resentment.

## BODY BLOW

I was undefeated but missed the world championships because I was hurt. So despite my record, I was still unknown. When it came to the Olympic trials I did well and made the team. Then I hurt my knee and had to have knee surgery. I had cartilage removed and after the surgery the doctors were not optimistic that I could make it back for the Olympics in five weeks. I got into fighter mode and I was determined to defy the doctors and make myself ready. Also, Bela has this magical power to make someone believe in themselves and he said to me one time, "You can do this," and I was off!

Not only did I make it back, I was even stronger mentally and physically after having gone through this intense emotional stretch.

## PROBLEM

The U.S. gymnastics community didn't exactly embrace me because I was not the stereotypical petite gymnast. There weren't many strong, powerful, and explosive women back then. I had Earl Campbell legs and was described as a fireplug and many, including the judges in America, did not like that image for the sport. I would think, "Okay, you don't like me, but I'm good and I'll win anyway." I think I broke barriers for a lot of young girls because of that.

*Wanna hear her schedule leading into the Olympics?*

Monday–Friday
7–11 AM training
12–3 school
5–9 back to gym

*Ouch!*

I didn't love every day, but I wanted that USA jacket, and the goal of the Olympics was so strong that I would not let fatigue or boredom creep in. Sure, I may have not walked the malls and gone to movies, but I did get to see the Great Wall of China, Europe, and much more.

## IN PERSPECTIVE

I know it worked out for me and I got the medals and the endorsements, but I also knew I was going to be okay in the world after I retired because of what I had learned in sports. I knew that the sacrifice, focus, and discipline would help me in my life and in any career I would choose.

## MY WRAP

*What a story! What a work ethic, and what a great person. You can see there were three or four times when she could have quit and few would have condemned her. I know I wouldn't have. Yes, she sacrificed part of her childhood, but then nothing about Mary Lou is typical. I'll tell you this much, my daughters will read her story first!*

# MARK BRUNELL

★ JACKSONVILLE JAGUARS CAREER PASSING RECORD: 25,689 TOTAL YARDS

★ SELECTED TO 3 PRO BOWLS; 1997 PRO BOWL MVP

★ NFL QUARTERBACK, GREEN BAY PACKERS, JACKSONVILLE JAGUARS, AND WASHINGTON REDSKINS, 1993–PRESENT

★ PLAYED IN 3 ROSE BOWLS AT THE UNIVERSITY OF WASHINGTON

The most interesting thing about this sport, at least to me, is the activity of preparation—any aspect of preparing for the games. The thrill isn't in the winning, it's in the doing.

—CHUCK NOLL,
*NFL Hall of Fame coach, Pittsburgh Steelers*

T-ball and football were my first organized sports, but like most kids, we were playing some sport all year round. My dad, who was a baseball player himself, was always the coach.

I didn't always stand out as an athlete. I wasn't big enough to play fullback, quarterback, or linebacker, so baseball came easier for me until football took over around the eighth grade.

The biggest impact sports had on me was always having someone there to encourage me or to set me straight if I had a bad attitude. If it was my dad coaching, he could also tell if I got overconfident or really down on myself, and he'd try to fix things, like a dad should.

*Wait a second. Mark Brunell, even a young Mark Brunell, had a bad attitude?*

### SACRIFICE

There were times when my friends would be going away on vacation and I would be home playing summer baseball or spring football and I'd want to do something else. But my dad always said that if you start something you'd better finish it. This really had an impact on me, and to this day I still do not quit anything, whether it's a game, a season, or a job. You have to play hard and stick with it.

If I were to walk into the locker room to look for whose dads still come to games, the number would be low. But I still talk to my dad twice a week and love it when he comes to games. Afterward, we always talk and I get his thoughts about how it went. I know he's the reason I am in the NFL, but his greatest contribution to my life is teaching me how to be a good father. I hear myself saying what he says to me and my three boys— "You don't quit." Hopefully those principles are transferring to them.

### GOING PRO

My dad kept me grounded. He never said, "You're going pro," and even when I got drafted I didn't think I could do this for a living. Even today, I'm still not sure why I made it, but my faith is so important to me and I believe that everything happens for a reason. I also believe that if I was not able to play football or if I hadn't been athletically gifted, I would have found a way to be successful in another arena. My success helps me today by giving me a platform to allow me to help people.

## LESSONS LEARNED

Success in football is all about hard work and keeping a positive attitude at all times. In that way football is very much like life. You have good seasons in sports and you have bad seasons. You have good jobs and you have bad jobs. There are good coaches and bosses and there are bad coaches and bosses. I've been able to take a lot of lessons in sports and then use them to help me handle life. This game is not life-or-death to me. My coach, Joe Gibbs, lectures us all the time that football should be fourth. The ranking in one's life should be: your relationship with God, family, friends, and then football. I know it's not as important as the first three, but that still doesn't stop me from pouring my heart onto the field every time I put on the helmet.

## PERSPECTIVE

Using that perspective on life has been very valuable to me. For example, 2004 was a miserable year for me. I was booed and then benched at FedEx Field and yet when I got home, my kids didn't care about that. They loved me just as much whether I won or lost.

## TIME TO TEST THAT BELIEF

I didn't know that soon after that experience I was forced to put my money where my mouth is. In 2005, it was Patrick Ramsey, not Mark Brunell, who was named as the starter. From the first time I watched from the bench. I refused to be mad or disgruntled. Sure, I had days when my attitude was bad. I went back to my faith and that philosophy of "everything happens for a reason," but I refused to believe I couldn't play anymore. This was the second real challenge in two years in my pro quarterback career. First, after nine years I was informed that my run at Jacksonville was over. They had a new coach and he wanted his own quarterback. It was difficult for me, of course, and it changed my plans, because I'd wanted to retire there. I felt tested, but I decided to make it exciting.

*This may be similar to what you experience at your job when the boss does not see you as the star you are or recognize the talent you have.*

The whole experience made me stronger. I don't know what I will do after football, but it's a reminder that there are tough times ahead of everyone. I'd like to think I will never go through times like that again, but I know that if and when the bad times do come, I'll be able to handle them because of what I went through the last two years. Tough times are a test and how I handle them is a story that's still being written.

## AFTER FOOTBALL

I'm pretty confident that without football I would have been okay. It's likely I would have been a coach, like my dad, and I would like to think I would be pretty successful doing that. Not in terms of wins, but in terms of having an impact on young people's lives, being a good role model. That would have been fine with me.

## HOW I PLAY THE GAME

I'd like to be remembered as the guy who played hard all the time and just didn't quit. I loved being the big underdog and having to find a way to win. The one game that I'll always remember was the Jaguars versus John Elway's Broncos in 1996. We beat them. If there is ever any doubt that I belong in this league playing at such a high level, I can always bring that game back to mind and I know I'll be okay.

## MY WRAP

*Mark didn't need football, but he loves the game. His career and its challenges mirror so many of yours even though you likely never did nor will put on the helmet and pads. He plays with passion, but he doesn't and never did let the game consume him, which is why almost all who know him say his life has true balance. However, few things would bring him more pleasure than a Super Bowl ring. With the Redskins, he might just have the team to do it.*

*Player, left, with Arnold Palmer*

# GARY PLAYER

★ DESIGNED OVER 200 GOLF COURSES AROUND THE WORLD

★ ONE OF ONLY FIVE PLAYERS TO WIN GOLF'S CAREER GRAND SLAM

★ 9 MAJOR CHAMPIONSHIPS (3 MASTERS, 3 BRITISH OPENS, 2 PGAS, 1 U.S. OPEN)

★ 24 PGA TOUR VICTORIES; 19 SENIOR TOUR VICTORIES; 120 OTHERS

★ PGA TOUR GOLFER, 1957–1984; SENIOR TOUR, 1985–PRESENT

When you play for fun, it's fun. But when you play golf for a living, it's a game of sorrows. You're never happy.

—GARY PLAYER

When my dad first asked me to play golf with him, I told him, "No, it's a sissy game." But he kept on me, so I grabbed a set of clubs and went out on the course. I struggled through that round, but I was hooked.

At the age of fourteen, I met a girl on the course whose father was the teaching pro and whose brother made it on the PGA Tour. I liked her and so I became instantly immersed in the game. Eventually I married her, and we've been together ever since.

My greatest asset on the course is my attitude. I always look for the positive in any situation. I can't control everything that happens to me, but 90 percent of the time I can control how I react to it.

When I started playing golf and began to think I could actually make a living at it, I was so thankful. I had a grateful heart then and a grateful heart now.

## CLIMBING THE LADDER OF ADVERSITY

I adored my mom, but she died when I was eight years old. Later, I found out that her dying wish for me was to go to a school that demanded a uniform and discipline. My dad made that happen. He was working in the South African gold mines, twelve thousand feet underground, but he earned a hundred bucks a month. My older brother was fighting in World War II. To get to this school, I had to travel ninety minutes via streetcar, and when I got home it was usually to an empty house. It was a struggle for me, and as I rode that trolley each day I swore that I would make something of myself and whatever I did, I would try with everything I was worth. I also came to believe that through adversity one climbs the ladder. I started to look at adversity as the greatest thing that could happen to me.

Every one of us will encounter adversity along the way, and we should warn our kids to expect it at some time in their life. If you don't warn someone about the way life works, then when they encounter trouble they're ill prepared and end up as alcoholics, committing suicide, or overeating, simply because they just can't handle it.

## SEEING THE WORLD

Traveling around the world has been a blessing in my life. Go to China and see people travel to work on their bikes. Travel through Africa and you'll get a perspective on how much we have. One week I've been able to have lunch with the President of the United States and the next week I'm in an African village eating with my hands. That's why I formed the Gary Player Foundation, to help educate the world with an appreciation of other cultures.

## THE BLACK KNIGHT

I modeled my life and my beliefs after my dad. Here's a man who had to leave school after only five years to feed his family, because his father had died. The only job he could get was in the gold mines, yet he spoke three black languages fluently, along with Dutch and English. He drilled into me, "Manners maketh the man." When I told him I was going to be a pro golfer, he kind of chuckled and said, "If you do this, come up with a brand." So I came up with the idea of dressing in black and being known as the Black Knight of the Fairway.

My dad also helped me get over any issues I had with my stature. He said, "In the mines, we use dynamite and it's very small but very powerful. Don't let people tell you you have to be big, because if you have that explosiveness inside you, you're tough. And Gary, you've got it."

## WORK ETHIC

I was and am an animal. I am a workaholic in life and a workaholic at my job. No man has ever hit more balls then I have. For example, two days ago (before this interview) I had cataracts taken out, and this morning I was playing golf. I just came back from an hour in the gym and will hit balls again later today. I'll also read and work on my speechmaking, because I address businesses throughout the year.

## DON'T TAKE ANYTHING FOR GRANTED

In 1962, I was playing with Arnold Palmer at the Masters in Augusta. I had him by two shots with three holes to go. I thought I was going to be the first man ever to win the Masters twice in a row. I was already ten feet from the cup on the sixteenth when Palmer hits the worst-looking shot. I said to my caddy, "I've got him. The best he can do from that spot is a bogey and the worst I can do is par." Wrong. He hit an impossible shot into the cup and I missed.

He won the Masters in a playoff and I said I would never again come to a conclusion before it was over. I set myself up for that letdown with poor thinking.

## APARTHEID AND ME

One day I walked up to my prime minister, who was a staunch believer in apartheid (a political system in South Africa that separated the black people living there from the white people and gave particular privileges to those of European descent) and said, "I want to bring Lee Elder (an African-American golfer) to South Africa to play a tournament." To my surprise, he said yes. It was the beginning of the official breakdown of apartheid in South Africa. Soon after, some people called me a traitor, but that was okay. There were demonstrations everywhere I played, and practically every day I had people say they were going to kill me. Even in America, I was hit with ice and had telephone books thrown at my back. As a result, I learned mental toughness, so it actually turned out all right for me.

## HOW TO LOSE

I don't believe the commonly held belief that goes, "Show me a man that's a good loser and I will show you a loser." That is hogwash! It's only a game, it's not your life. Let me tell you how it's helped me in my life. I put a million bucks down for the first option on a piece of real estate. Well, a friend of mine backed out on the deal, so I took him to court. The court ruled on his behalf and I lost a million dollars. Afterward, I walked up to him and said, "Tim, you beat me. Congratulations."

He was stunned and at first he couldn't even speak. Then he said, "I can't believe you're congratulating me after losing a million bucks."

I said, "You beat me and that's that." And we're still friends today.

**MY WRAP**

*The lessons in Gary Player's life are too numerous even to recap. He spent the first half of his life in a societal hurricane. He learned to embrace the adversity, and his life has become so much more than simply about golf. His new goal is to help fifty million youngsters, and don't you dare doubt that he'll make it happen.*

# CAL RIPKEN JR.

★ **BASEBALL HALL OF FAME, 2007**

★ **PLAYED IN A RECORD 2,632 CONSECUTIVE GAMES**

★ **19-TIME AL ALL-STAR**

★ **TWO-TIME AL MVP, 1983, 1991**

★ **WORLD SERIES CHAMPION, 1983**

★ **AL ROOKIE OF THE YEAR, 1982**

★ **SHORTSTOP AND 3RD BASEMAN, BALTIMORE ORIOLES, 1981–2001**

Give a boy a bat and a ball and a place to play and you'll have a good citizen.

—JOE MCCARTHY,
*Hall of Fame baseball manager*

I always wanted to be in the thick of things, and that's why I played center half in soccer. From that position, I could control the game and see the entire field. And that's the same reason why I chose to play shortstop in baseball.

## THE MOMENT

I made the varsity team as a freshman, almost by default because we had no one else. I couldn't even throw the ball from shortstop over to first base, so I played second. I hadn't grown yet, I was just 5'7", 128 pounds. And why do I remember my dimensions? Because we had to weigh in in front of the whole team and when I did I heard snickers and laughing as I stood there.

*He'd grow to 6'4" of solid steel.*

It was humiliating. I thought about that for a long time, because most of the other kids were seniors and they practically had full-grown beards. Now, on any team I've ever been on, I never looked at anyone as an outsider. To this day, I hate rookie hazing. Fortunately, I didn't have to go through that once I reached the pros, because we had a mature, professional team that had the right attitude. As time wore on, though, the team changed and they used to do things like make guys dress up in women's clothing. I never approved of that kind of behavior and if I could have I would have stopped it.

I've always been sensitive to other people, and especially when you're trying to build team chemistry, I don't believe you have to break someone down in order to build them back up. I think you should build them up and bring them into the fold, and I've always gone out of my way to make new teammates feel welcome.

## NO STAR STATUS FOR HIM

We live in an age where some people have the attitude that they're too good for someone else. I call that "Big Leaguing it." For me, everyone is equal. I know I might be a starting player and I might be making more of a contribution, but that doesn't mean everyone else is less important than I am. When I first got called up to the majors, I had my at bats and then got sent down to Triple-A because in taking batting practice I hurt my shoulder, so I couldn't continue playing. Some of the guys on the team thought I was dogging it and they treated me pretty poorly. Eventually I recovered from the injury, starting hitting, and led the league in home runs and RBIs, and suddenly their attitude toward me changed. As a result, I saw both sides of the game in one season and in doing so decided that I just didn't want to make anyone feel like a second-class citizen. My problem was that I was also shy and it took longer for me than my outgoing brother Billy to get to know the other players. Despite how it might have seemed, I always had compassion for the outsider.

### KEEPING THINGS IN PERSPECTIVE

My dad, as you may know, was a baseball player, coach, and manager for his entire life. He would never let our heads get too big because we could hit, throw, or catch. He'd say, "Everyone has his or her individual talents. You may be good at the game, but it's nothing without the fans." Being with the fans, and signing autographs as often as I do, is my way of keeping them involved.

### MY DAD AS MY DAD

Every time my dad would leave on a trip he would pull me aside and say, "Cal, you're the man of the house now and you have to take care of the family." He'd shake hands with me and just turn around and leave. It would be a kind of a formal thing. And so, even at the young age of eight, my dad made me feel part of the structure of the family.

### ALWAYS A "WHY" KID

I was a shy kid in social settings but not in a baseball setting. I would watch my dad work with young players, taking it all in, and I'd always be asking him why he did what he did after just about everything. He'd always say, "If you want to know about something, go to the source." And that was, more often than not, him.

## FACING THE FEAR

Goose Gossage was a big, intimidating stopper and I was not looking forward to facing him. I was actually scared out of my mind because I saw him hit Ron Cey, the Dodger's third baseman, in the head with a 100 mph pitch in the World Series and not care that he did. The first time I faced him I wanted to step out of the batter's box. Well, after going 0 for 8 against him, I realized I had to solve my issue with this guy if I was ever going to stand in against him.

One night, when he was in Baltimore, I heard he was going to be at a bar right by my house. I just happened to drop in and he called me over to his table and I quickly learned he was a nice guy, not a monster. It took the edge away and soon I got some hits off him and I no longer felt like jumping out of the box. I learned that if I couldn't hit someone or if I was intimidated by a guy, I should go befriend him and humanize him. It helped me on the field over and over again, and has also helped me with relationship issues off the field.

## IRON MAN

My dad was the hardest worker I know and I wanted to be like him. I learned it was my job to show up at the baseball game and put myself in the hands of the manager. It wasn't my job to tell him when I could or couldn't play. It was up to him. The cool part for me, as I approached Gehrig's record, was that I got letters from people all around the country who had their own streak, like those who hadn't missed a day of work in thirty-four years. There is a pride in having the courage to show up and take part in something every day, and that's what came out of all those stories. The letters showed me people were proud to accomplish something rather than taking the easy way out. And there is an easy way out and that's to take a sick day, or go into the manager's office and say, "I need a day off." The hard part is trying to meet the challenge. Believe me, it would have been easy for me to take the day off if I was 0 for 25 and Roger Clemens was pitching the next day. As my dad would say, "You have a game, you can play. Let the manager know you want to play."

The principles I learned were not just for baseball, they were for life.

When you play sports, no matter what the level, these are the kinds of principles that should be taught early on. The lessons learned in sports can and should be used later in life. For instance, instead of a game, you have a speech to prepare for. You don't wing it, because in sports you learn that failure comes when you fail to prepare.

## HAVING A PLAN

My dad let me know I was likely to be drafted right out of high school, but we had a plan. Before I signed I wanted to make sure that if it didn't work out, I would get my college tuition paid for and I could start my life at twenty-six. The odds were against anyone making it and doing well, and I just wanted to have a safety net in case I washed out.

*It sure did work out for Cal.*

## RETIREMENT? YEAH, RIGHT

I left a job that made me put a uniform on with my name on the back of it and now I have a uniform that doubles as a suit. I still have challenges and issues in front of me and I try to accomplish things the only way I know how, which is to put my head down and work hard to solve the problem. It's something that comes naturally to me, because I played the game the same way. I bought a couple of minor league teams (the Augusta Green Jackets and the Aberdeen Iron Birds). I'm looking at a few others. I've built two youth complexes in Myrtle Beach. I'm an author and I'm working on another book. I formed a little league, which is a division of the Babe Ruth league called Ripken Baseball. I would not have been able to do all these things if I didn't have the organizational drive that I learned playing baseball.

## FINAL THOUGHTS

Have the courage to give it a try, no matter what it is. You can't succeed unless you're willing to fail. I played basketball just as hard as I played baseball, with only a fraction of the success. I didn't care, because my end goal was to get in great shape and I was not willing to play it safe. I didn't ease up to protect my consecutive games streak. I believe that you're always the safest when you play the hardest.

**MY WRAP**

*He's the hall of famer who may be the most comfortable in the grass roots of the game, with a blue-collar work ethic and a personality with blue-chip stats. He might just have the greatest work ethic in sports history. Look at what he does now and tell me sports did not teach him how to set and accomplish anything he put his mind to. Cal also knows his hall-of-fame parents set him up for a hall-of-fame life.*

MARION MORRISON - JOHN WAYNE
*The Duke* - CLASS OF 1925
GUARD - WEIGHT 170 - 2 YEARS VARSITY

# JOHN WAYNE

★ **DEPICTED ON U.S. POSTAGE STAMP, 2004**

★ **ACADEMY AWARD FOR BEST ACTOR, *TRUE GRIT*, 1969**

★ **LEGENDARY FILM ACTOR, DIRECTOR, AND PRODUCER; APPEARED IN OVER 200 FILMS, 1926–1979**

Breaks balance out. The sun don't shine on the same ol' dog's rear end every day.

—DARRELL ROYAL,
*longtime University of Texas football coach*

CONTRIBUTIONS FROM DON SHOEMAKER, FORMER GLENDALE HIGH SCHOOL HEAD FOOTBALL COACH; GRETCHEN WAYNE, DAUGHTER-IN-LAW; MICHAEL WAYNE, SON; PATRICK LANCASTER, HISTORIAN; AND CRAIG FERTIG, UNIVERSITY OF SOUTHERN CALIFORNIA HISTORIAN.

DON SHOEMAKER:

When I was coaching high school football, John Wayne (born Marion Morrison) meant a lot to my team because he was a marquee actor. Even today the kids know him, and they see the pictures in the football program that we hand out every year.

Wayne played football for three years. In 1920, at 6'4", he was like a giant. If USC was even considering him, he must have been a heck of a player. I talked to kids about what John Wayne did here not only as a player but as a student. He was elected president of his class, he wrote for the student newspaper, and he started on the football team. Every parent thinks their kid has to get a scholarship and that to do that he or she must focus on one thing. Wayne did it all, and he ended up being an American icon. Also, keep in mind that this future megastar was both an offensive and defensive lineman, positions that got little credit then and today. He was a leader who didn't need the spotlight, but it kind of found him.

PAT LANCASTER, HISTORIAN:

By many accounts, Marion Morrison was a shy, withdrawn boy growing up.

His experiences in high school, where he blossomed, changed all of that. Athletics played a huge role in that metamorphosis. As a freshman, he played on the lightweight football team and looked to be not much more than one hundred pounds. He wasn't involved in many other activities in school, but as his athletic abilities grew, so did his confidence, which you can see by looking through the yearbooks from 1922 to 1925.

By his senior year, he was a strapping, good-looking boy, listed at one hundred seventy pounds. What was of note was not that had he been starting on the varsity football team for two years, but that he was in numerous clubs. In the team picture, Morrison is featured front and center, holding a football.

In 1924–25, it seemed that he was a leader in everything at school. He was senior class president, sports editor of the school newspaper, *Explosion*, chairman of the senior dance, and winner of a bronze pin for academic excellence. His classmates trusted him to lead the senior class. His football teammates counted on him as he led them to the Southern California championship. And from there, his life and career really took off.

I wonder what it must have been like to be playing football next to Duke Morrison. His leadership, his strength, and his true grit, even at a young age, surely must have rubbed off on his teammates.

A story that his son Michael told us while addressing the crowd at Glendale High School's one hundred-year reunion in 2001 was that he, Michael, was a student at Loyola, a private high school in Los Angeles, and they had a football game scheduled against Glendale High one Friday night. It was in the early 1950s and the game was being played at GHS. He said his dad sat on the Loyola side for half of the game and then, showing his loyalty to his alma mater, he went and sat in the stands on the Glendale High side for the other half. His presence on the GHS side caused a bit of a stir, but it was certainly a thrill for the GHS students, teachers, and alumni.

Michael said that he felt it was important for him to participate in Glendale High's one hundred-year anniversary because it had been such a special place to his father.

## TYPE OF PLAYER

While it's tough to find someone who saw Marion Morrison play football, his school newspaper accounts indicate that he was special. December 5, 1924, as it appeared in *The Explosion:*

> GLENDALE AND ORANGE PLAYED TO 0-0 TIE
> MORRISON PHILLIPS AND DOTSON
>
> The line was the same famous stonewall. Each man did his full share. Heincke touted as the best player on the Orange team did not show so well against Morrison. Well, you can't blame him.

Later the same season:

## GLENDALE TIES VAN NUYS 6–6
## MORRISON PLAYS A STELLAR GAME

Marion Morrison, the husky left guard, was the outstanding lineman. Time and time again he would nab the Van Nuys men behind their own line and he was a terror to the several guards and tackles that faced him.

### COLLEGE-BOUND

According to his son, Patrick Wayne, football meant a lot to his dad. Most importantly, it got him a scholarship to USC, without which he could not have attended college. However, his dad suffered a setback when he injured his shoulder after his freshman year. This meant that he couldn't play sophomore year, and he lost his scholarship. Had he not hurt his shoulder, he would have continued his education and it's unlikely he would have gone into the movies, but according to Patrick, he might have been President of the United States.

CRAIG FERTIG, USC HISTORIAN:

Even though he never got his degree from USC, John Wayne never lost touch with the school. One day, legendary coach John McKay called on Wayne to address the team before they were to play Texas in 1966. According to Fertig, Wayne's impassioned address made Vince Lombardi sound like a kindergarten teacher. USC went out and beat the favored Texas squad 10–6. The next week, from his movie set in Mexico, he wrote McKay a one-page letter, and here's an excerpt I love.

> I think the Texas bunch were good losers, but tell the squad we don't care whether their opponents are good losers or bad losers as long as they lose.

### MY WRAP

*From all I've read, John Wayne would have been a standout without football, but certainly he got additional attention and respect from his exploits on the field, and he sure took pride in having played, according to his family. He came from the humblest beginnings, defied the odds, and overcame a huge set-back when his shoulder injury caused him to drop out of school. He jumped into film as a prop guy and wound up being the ultimate symbol of the tough, rugged American. Not bad. I think he perfectly described what was underneath the gritty confidence he brought to every movie he was in and to his life when he said, "Courage is being scared to death, but saddling up anyway." Maybe he was scared when he lost his full ride at USC, but he moved on, and his life was one big high right up until his passing in 1979.*

# REBECCA LOBO

★ **COLOR ANALYST FOR NBA-TV**

★ **WNBA CENTER, NEW YORK LIBERTY, HOUSTON COMETS, AND CONNECTICUT SUN, 1997–2003**

★ **OLYMPIC GOLD MEDAL, 1996**

★ **NCAA CHAMPIONSHIP, 1995**

★ **NAISMITH COLLEGE PLAYER OF THE YEAR AWARD, 1995**

I've always wanted to equalize things for us . . . Women can be great athletes. And I think we'll find in the next decade that women athletes will finally get the attention they deserve.

—BILLIE JEAN KING,
*legendary women's tennis player*

I had a brother who was six years older and I was always the tagalong little sister who wanted to do what he was doing. I especially loved basketball, and I played all the time, nonstop dribbling and shooting. And because I was more skilled and was exceptionally tall, I had success right away.

## SUCCESS FROM THE START?

If I was to cite the key factor in my athletic success, it had to be my parents. My dad was a coach, but never coached me, and my mom was never critical—she was always supportive. Anytime my grades would slip they would threaten to take basketball away from me, and that would straighten me right out. It was great to know that win or lose, or no matter how much I struggled, they would treat me the same. Best of all, they never yelled at the refs, which would have been embarrassing and set a bad example.

## BIG GAME. BIG HOPES

It never even sunk in that I could play basketball in college until I got to seventh grade. My brother had just gotten recruited and was committed to play at Dartmouth, so, once again trying to follow in his footsteps, I went to basketball camp at Dartmouth, and while I was there I can remember a coach telling my parents, "You think your son was heavily recruited, but your daughter will be much more sought after." Hearing this realigned my thinking, because prior to that I was just hoping to play in college, having no idea that I could get a scholarship or that I was good enough to be recruited. As a kid, I had a poster of the women's Olympic team and that was a goal no one talked about—women's college hoops. After all, there was just one TV game on a year—and that was the final, and it was usually delivered tape-delayed.

## PICKING A COLLEGE

The first time I made a big decision that was against my parents' wishes was when I chose the University of Connecticut as my college. I had offers from Stanford, Northwestern, and Notre Dame, but I liked Connecticut and especially the coach, Geno Auriemma.

## MY REALITY SHOW

From day one, I found out college ball was a whole new ball game. I thought I was in good shape, but I wasn't. I thought I knew how to work hard, but I didn't. I had to learn how to push my body to the limit. One time, we all went to the track for a timed mile and the coach was unhappy with me and some other players and so he just ripped into us. I was in shock, because no one had ever talked to me, or anyone around me, like that.

## BIG MISTAKE

I did not have an easy time. My first two years I got yelled at every single day by Coach Auriemma. He would call me the worst post player in the country . . . or worse. I called home quite often, until finally my mom said, "You have two choices. Either you transfer or you go in there and talk to him. This is your problem, not mine, and you have to handle it. We will not fight your battles, so make a choice."

## TURNING POINT

So I went to see Geno and asked him what his problem was, why he was yelling at me every day. And I told him that I wasn't happy. He proceeded to reach into his drawer and pulled out our media guide. He opened it to my bio and read down and said, "Is your dream still to be on the Olympic team?" I said yes and he said, "Well, it's my job to make this happen and until you're giving me a great effort every day in practice, I'll keep riding you."

What he said opened up my eyes as to why this crazy man was all over me every day, all the time.

*I know you're thinking that the coach could have told her why he was rupturing his spleen every day in practice over her and saved Rebecca two years of anguish, but she explained she was not ready to have the conversation until she initiated it.*

## PAYOFF

We still had some tough moments, but after that talk I really began to understand what it took to be successful and what it meant to work hard. My junior and senior year I was an All-American and in 1995 we won the national championship. It all worked out, but I didn't have to win the title to feel like it was all worth it. I learned to push myself and demand more of myself and still do to this day in everything I do.

## THE OLYMPICS

I was able to take the momentum from the season and fulfill my childhood dream of making the Olympic team. What made it even better for me was winning the gold and doing it at home in Atlanta at the 1996 games. The year before, I was cut from the national team, but after my senior season I improved enough to make the roster. I ended up playing two years in WNBA but after two years called it quit due to injuries.

## HOW DOES IT ALL HELP HER TODAY?

When I was complaining to my parents a lot and they said that it was my problem, that I had to solve it and I did, the lesson did not end there. To this day, anytime I have an issue, whether it's a speech or a problem while working for ESPN or a problem with the family, I don't wait for anyone else to step forward, I try to solve it myself, head-on. I don't think I would be like that if I didn't have early turmoil and was pushed to tackle my troubles myself.

*Thanks, Coach.*

## MY WRAP

*Rebecca was not only blessed with talent and parents who believed in her, but also with a great role model in her older brother. He played hard and fair and was a tremendous teammate. She did the same because she wanted to be just like him. This should be a reminder to all of us with younger siblings: they're always watching us!*

# GEORGE FOREMAN

★ 2-TIME WORLD HEAVYWEIGHT BOXING CHAMPION, 1973, 1994

★ FOUGHT MUHAMMAD ALI IN ZAIRE IN WHAT BECAME KNOWN AS
   *THE RUMBLE IN THE JUNGLE,* SUFFERING HIS FIRST DEFEAT AS A
   PROFESSIONAL BOXER

★ OLYMPIC GOLD MEDAL, HEAVYWEIGHT, 1968

Losing is the great American sin.

—JOHN R. TUNIS,
*writer*

Playing football at the Job Core Center was the first time sports started playing a positive role in my life. I had to show up on time, practice hard, and, more importantly, I was separated from all these guys with bad habits. I knew then that if I stuck with sports, I might have a future. It was also the first time I did anything as a team. Most teamwork first comes at home with mother, father, brothers, and sisters. Well, my father left early and my mom had tuberculosis, so she was in and out of the hospital and I was often on my own. When I joined a football team at seventeen, it was the first time I felt a part of something—and the first time I felt anything positive.

### THE MOMENT—THE MAN

At one practice, the head coach saw me line up and get into stance and noticed that my fingers were open and out. He said, "George, put your fingers in or you'll get one of them broke." I was stunned. This guy actually cared about me. It sounds like a little thing, but it really meant a lot to me. He told me something I did not know and just the way he said it, well, he had me for life. As far as I was concerned, I'd go through a wall for him. Too bad the boxing coach, Charles "Doc" Broadus, saw me soon after that and disrupted my plans of playing college football. Doc stayed on me. He got me a license and took me to spar with Sonny Liston. He used to say to me that I could be an Olympic champion, but that I'd have to stop smoking, drinking, and street fighting. And you know what? I stopped. I just wanted a goal and someone to take an interest in me. And he did both.

### MOMENTUM

Sports saved my life. Sports made me want to read and go back and get my GED and, in turn, gave me a future. When I won the Olympic gold medal in Mexico City, I thought I was done with sports and boxing. It didn't turn out that way. I was overwhelmed with offers to go pro, so I did, and I became a two-time heavyweight champ.

## HOW I PLAY THE GAME

It's all about preparation, putting in the work in order to get something out of it. I never was the best boxer or the hardest puncher . . .

*Gee, George, I don't necessarily agree with that . . . nor would some of your opponents, I'll bet.*

. . . but I gave it more time than anyone else in the world. I worked hard preparing for a fight, digging holes, chopping down trees. That's how I had the endurance to fight so often, even from the ages of thirty-nine to forty-five.

## DISILLUSIONED

Before the Ali fight, I was feared in the ring. People all around me were telling me how great I was, how invincible I had become. Once I lost to Ali, all those people vanished. They stopped returning my calls. I was left in my locker room with just my dog, and no world title. It was a very sobering experience.

## A SECOND CHANCE

When I came back after taking ten years off, I was determined to enjoy myself. When I had beaten Joe Frazier in Jamaica, I never even saw the water. And I didn't much like the media. In Zaire, I never left the gym. This time, I made it a point to see every city I fought in. And I was determined to be friendlier with the press. I wanted to make sure that, for the first time in my life, I would enjoy the process. I collected all the headlines written about me and now I have the scrapbooks to help me look back on those times. When someone says they saw one of my old fights, I say thank you and answer their questions rather then walk by them. At an age when most say you can't change, I learned how to change and now have had a great life because I made those changes. I took the same new me into broadcasting.

## ON NEEDING FEAR

People think they have to ignore fear to get rid of it. Over the years, I learned to embrace it. I was not scared in Zaire against Ali and I was knocked out. Now I embrace fear and it motivates me. I use it to keep myself strong. I feel that same kind of fear when I preach at church or speak in front of a big audience. I fear I won't have a clear message to all those people who have given me their time. And yet, if I have that fear, I am optimistic that I will do well.

## MY WRAP

*Everyone knows George Foreman, but few know how interested he is in other people and how grateful he is for his success. I get the feeling that there is going to be yet another chapter to his incredible career, although I'm pretty sure it won't be in the ring. George might be the most clear-cut example of the many ways sports shape and often save lives.*

# PETE CARROLL

★ LED USC TO 2 STRAIGHT NATIONAL TITLES

★ PAC-10 COACH OF THE YEAR, 2003–05

★ FOOTBALL COACH, 1974–PRESENT, INCLUDING HEAD COACH OF
  THE NEW YORK JETS, 1994, AND NEW ENGLAND PATRIOTS, 1997–99

★ HEAD COACH, UNIVERSITY OF SOUTHERN CALIFORNIA,
  2001–PRESENT

Experience tells you what to do; confidence allows you to
do it.

—STAN SMITH,
*Hall of Fame tennis player*

## ORANGE BOWL: CHARACTER MATTERS

Working with head coach Lou Holtz at Arkansas was a tremendous experience. One year we were prepping for the Orange Bowl and a few of our best players violated team rules. He cut all three of them, even though we were playing an incredibly powerful Oklahoma team who we'd have trouble beating even if we were at full strength. We ended up killing Oklahoma because the team came together when Coach Holtz took a stand. In other words, the team we put on the field that day became more powerful, even though they were less talented.

## THE MOMENT

I was having success as a coach but not as a head coach, and after one year with the Jets and three with the Pats, I found myself at home without a job, wondering what I should and could do next. Someone handed me John Wooden's autobiography and in it I discovered that the greatest coach in sports history needed time to map out and write up his own belief system and philosophy. What I took from Wooden was his belief that you never talk about winning, whether it's a game or a title. It's all about doing the things needed to win, putting in the effort, and staying with a program. Our goal at USC is to do things better than they've ever been done before, and what has happened in that effort are many wins and two national titles. What we are striving for is a competitive edge, and at the same time I also try to build fun around it, which, by the way, is how I live my life.

## YOU CAN'T DO IT WELL IF YOU DON'T LIKE IT

Most people do not know what they believe in and most people do not know what their approach is, and I am not just talking about coaches. I am talking about everyone. Most people don't take time to figure it all out. During most of my career I was guilty of that, too. It's like going on a trip without knowing your destination. You won't know how to get there or even how to recognize it when you do arrive. I was close to making it all work for the Patriots, but I just didn't have the power. Too many people were in charge there. At USC, they just leave me alone. Thankfully, it's all come together.

## GIVING IT YOUR ALL, BUT NOT GETTING ANYTHING

I was a real good athlete. In fact, I was the best at almost every sport I tried until I got to high school. I just stopped growing. I tried to play football at 115 pounds and needed a doctor's note just to be able to try out. It wasn't until I got to college at the University of the Pacific that I matured. I went a long time without having any substantial success, but I just hung onto the dream of playing college football and maybe getting a shot at the NFL. I developed a kind of chip on my shoulder. I knew what I was capable of, and only my family understood what I was going through. But what I went through has helped me become a better coach. I understand the guys on the bench in the same way I understand the stars, because I feel like I've experienced both roles on a team. I wound up having a lot of success at Pacific, and even tried out for the World Football League.

## FINALLY

I had to go through those hard times with the Jets and Patriots to become the coach I am today.

*Fired after one year with the Jets? Now that's tough!*

I hate learning the hard way, but I did, and I am enjoying every moment I am experiencing now at USC because of it.

## MY WRAP

*Guys like Pete Carroll, from his days as an undersized athlete to his raw deals in the NFL, never quit. Pete used his downtime to get better, to collect more information, and now he has built a dynasty at USC. Could another NFL stint be in the cards?*

# TYRONE "MUGGSY" BOGUES

★ COACH OF THE WNBA CHARLOTTE STING

★ SHORTEST PLAYER (5'3") EVER TO PLAY IN THE NBA

★ NBA POINT GUARD, WASHINGTON BULLETS, CHARLOTTE HORNETS, GOLDEN STATE WARRIORS, TORONTO RAPTORS, 1987–2001

If you don't make a total commitment to whatever you're doing, then you start looking to bail out the first time the boat starts leaking. It's tough enough getting that boat to shore with everybody rowing, let alone when a guy stands up and starts putting his life jacket on.

—LOU HOLTZ,
*legendary college football coach*

I started really caring about the game when I was eight years old. I would travel to the city of Baltimore to play basketball. The summer is when all the real talent came out, and to get anywhere you had to find a way to make a name for yourself by the time you were twelve. I was always the smallest kid on the court, so I had to try twice as hard to carve out my own niche and prove people wrong.

### COULDN'T HAVE A BAD GAME

It didn't really bother me that I was the smallest kid in every game I played. However, because of my height, I felt I could not have a bad game. I looked at each game as a chance to prove myself and disprove what some might think I could or couldn't do.

### YOU'RE GOOD ENOUGH, BUT NOT GOOD ENOUGH TO MAKE IT . . .

People would talk about how well I could play, but they were also quick in trying to limit me in what they thought I could do. I was a star at Dunbar High School, the best team in the country, but I would still hear, no way would I go to a top Division I school. Well, I went to and started for Wake Forest for four years in the mighty ACC. I would hear how good I was, but I was often told I would never go pro at 5'3". I was drafted in the first round by the Hornets and played ten years before being traded. What they didn't know is, I *had* to make it. I lived in the projects and it was a way out.

### THE ONE GAME

I remember playing Camden, which was, at the time, the number-one high school team in the nation—we were number two. They had the top-rated point guard in the country, Kevin Walls, and I knew I would have my hands full. Then it happened. Right before tip-off, he walked up next to me and pointed down to my head (to highlight how easy a time he expected to have against me because of my height). The crowd laughed, and I think the whole building was staring at me. We went back to the huddle after the first time-out and the coach pulled me over and asked if I was all right and I just said, "Coach, this is a party and we are here to take care of business." We did. We went in at the half up thirty-one points, eventually winning by twenty-five. How did I do? Seventeen points, eight steals, eight assists. It was a game I still think about and talk about, even today.

## PROUDEST MOMENT

Earning my college degree is what I am most proud of. I didn't get my degree after four years at Wake Forest, but I went back and got it after joining the NBA. I just had to have it to show my kids that education matters. I wanted to do it for my mom, to reward her for sacrificing so much so I could play. It also helps me when I talk to kids about the need to go to college. They can relate more to me as a 5'3" NBA player than a 7'3" one because there are more kids my size than Kareem's or Shaq's size.

## KIDS

I remember what it was like growing up and I know how cruel kids can be to one another, so I spend time going to schools, trying to let them know there is always hope and a way out of any circumstance, no matter how bad it is. Most important, I let them know the need to set goals and overcome the naysayers who will tell them they can't do something even though they know they can. By the way, the kids who never saw me play more than likely have seen my only movie, *Space Jam*, so it's kinda nice that they enjoyed it, and it helps give me some credibility with them.

## MADE IT BEFORE THE NBA CAME CALLING

If I'd banged up my knee or didn't make it in the NBA, I think I still would have been okay because of what I went through. Getting to college and away from the street was what defied the odds. Going pro was what came next, and if that hadn't happen I would have found something else. Basketball was a treat; growing up where I grew up was the challenge. I saw where basketball could take me and rode that wave as long as I could.

## MY WRAP

*The odds against playing NBA basketball are long to say the least; the odds against playing basketball in the NBA at 5'3" are unimaginable. But it seems that the only person not impressed with Muggsy Bogues is Muggsy Bogues. His life is an example of using your strengths and not worrying about your weaknesses. For example, he may have been shorter but he was usually also quicker. He would work to get his opponent to play to his strength rather than their strenght, which is a lesson we all could use.*

# TONY STEWART

★ NASCAR NEXTEL CUP SERIES CHAMPION, 2005

★ FIRST AND ONLY DRIVER TO WIN CHAMPIONSHIPS IN STOCK CARS, INDY CARS, OPEN-WHEEL MIDGET, SPRING, AND SILVER CROWN CARS

Problems are the price you pay for progress.

— BRANCH RICKEY,
*baseball executive*

B y the time I was eight, while all my friends were circling around the block on their bikes, I was racing go-karts at fifty miles per hour, which made me the hero of the neighborhood. Even back then, I didn't like getting beaten. I always had that fire in me to win at everything, from board games to baseball. It was only about winning.

## BUILT *NOT* TO WIN

As a kid, I remember that my family went through some tough times financially, but we always managed to get through it.

My dad and I would be out there spending money on racing, and my mom and sister would often get the short end of the stick. Racing at any level is expensive, and if you did run on the cheap you almost never got a good result. Unfortunately, that's what happened to my dad and me. Back then, at that level, teams would go three races on new tires and then junk them. What we'd do is buy those junked tires from another team and then we'd use them. That's probably why I started doing poorly, which just got me more and more frustrated. Even worse, I started to doubt myself. What made it even tougher was knowing that I was racing on my family's money. They spent more than they could afford so I could race. When I lost, well, that would just compound the pain. It wasn't easy knowing your dad had to mortgage the house to buy you the go-kart, especially when you lost.

*There's emotional pain and there's physical pain, and it seems that young Tony had both. At eleven, his dad yanked him out of his go-kart by his collar because he didn't win a race. After another loss his dad tossed out every wrench in his toolbox. To many, it seemed that Tony was driving for Dad. I think the observation is 100 percent correct.*

Things got so stressful, especially between me and my dad, that we just took a year off from racing. I am not saying I ever quit the sport; it's just that we took a time-out, a pause, because of the tension between us. And we were basically out of money

*Tony won the world go-karting championship in 1987, National Midget*

*championship in 1994, NASCAR Rookie of the Year in 1996, and the Nextel Cup in 2002 and 2005. As much as people loved his racing, some had a problem with his quick temper. To some, he had become NASCAR's bad boy, and he didn't like having that label.*

### AN ATTITUDE ADJUSTMENT

I decided I was tired of constantly apologizing for my temper. I didn't like getting booed at the track, and after meeting with my crew, who really let me have it, I decided to change. So, I lined up some mentors—the late Dale Earnhardt, Joe Gibbs (owner of his racing team and coach of the Washington Redskins) and Bob Nardelli, the former CEO of Home Depot. Their leadership and guidance have been extremely valuable to me. Now I'm surrounded by a great team, from racing to marketing to media, and I like to think I'm finally built to win on and off the track.

### PROUD AND FORTUNATE

I'm proud of being a kid who barely finished high school and now has not only had success in racing, but also in business. I am so fortunate to be able to have the success racing that I have and I owe so much to my team, both on and off the track. They're the ones who've helped keep me on the track and kept my head screwed on straight.

### HOW HE PLAYS THE GAME

I've learned that although it's me racing, it's not *my* sport. I learned how to get along with NASCAR, and as a result, things have been much better for my team and for me. I don't understand everything they do, but I've learned to accept that they have a reason for doing it. It's a relief not to fight the system. I still race as if my parents own the car, I don't have a dime, and someone else's livelihood is at stake.

### MY WRAP

*Like many of you, Tony's issues did not end once childhood came to a close. His seemingly overbearing dad and mom divorced and he personally had an engagement fall apart. Through it all he tended to wear his emotions on his sleeve, but it's heartening to see the effort he put forth to change and right what he saw was wrong. Buying back his childhood home, fixing it up, and giving the garage over to his dad, who could finally set it up like he wanted to, was a step in that direction. He's not only getting to be a better person, but a better driver, which is scary news to the rest of the Nextel circuit.*

# GEORGE SHULTZ

★ **U.S. SECRETARY OF STATE, 1982–89**

★ **U.S. SECRETARY OF THE TREASURY, 1972–74**

★ **U.S. SECRETARY OF LABOR, 1969–70**

We all choke, and the man who says he doesn't choke is lying like hell.

—LEE TREVINO,
*Hall of Fame golfer*

When I played football it was very different from the way it's played today. Coaching from the sideline was prohibited, and when a player came out of the game he couldn't go back in until the next quarter started. Needless to say, I played offense and defense. I played blocking back and linebacker. Back then, it really was eleven versus eleven. Now it's one organization against another organization. Frankly, I will take the old game over the new game any day.

Sports taught me relentless accountability. I learned responsibility to myself and to my teammates. If I missed a block or a tackle, I was letting the whole team down. I never wanted to have that happen.

I started playing in high school, but I wasn't a star. At Princeton I got very beat up. I had my ankles broken and my knees were wrecked, much like most of us who chose to play football back then.

### MY YEAR, MY MOMENT

Senior year, I was in wonderful physical condition. Even the coaches were commenting that I was doing something special in every preseason scrimmage. But I got clipped and I wrenched my knee to the point where I was not able to play my entire senior year. What happened out of something bad was something very good. I got my first teaching job coaching the freshman backfield, and this was the beginning of a lifelong passion for teaching. I just loved working with people and getting the most out of them.

### FINAL THOUGHT

Looking back, I feel sports should be a part of every university education. I'm not necessarily talking about high-level sports, just sports in general. In the classroom, you address your intellect, but on the playing field it's about character and judgment, responsibility and accountability, and listening to your gut. A good education develops your mind and your gut, and it worked for me.

## MY WRAP

*Another successful businessman and politician who is incredibly humble and self-effacing. Like so many others in his position, he gets that it's not all about winning but about playing.*

# JULIE FOUDY

★ NATIONAL SOCCER HALL OF FAME, 2007

★ U.S. WOMEN'S NATIONAL SOCCER TEAM, 1987–2004; CAPTAIN, 2000–2004

★ OLYMPIC GOLD MEDAL, 1996, 2004

★ WORLD CUP CHAMPION, 1991, 1999

A 100 percent concern with a game to the exclusion of all else is surely tinged with obsession. The single-mindedness necessary to fight one's way to the top, in no matter what sport, is something not shared by the majority of mortals.

—PAUL GALLICO,
*sportswriter, novelist, author of* The Poseidon Adventure

I didn't start playing soccer on a team until I was seven, but a couple of years before that the neighborhood guys used to come over and get me to play because I was the only chick who wanted to be out there.

The game always mattered to me, but in a healthy way. It's not like I went into a deep depression if I lost on Saturday.

## HARD WORK, SWEET REWARDS

The coach of the first team I was on told us that to play on his squad we had to be able to juggle twenty times. I didn't know what he was talking about. *Juggling like a clown?* I thought. Well, I found out that what he meant was keeping the ball up in the air, so I went into my backyard and taught myself how to do it. I just loved juggling and teaching myself new moves. I did it because it was fun and also because my club gave out awards for accomplishments. One hundred juggles got you a patch; fifty juggles got you a sundae. I just had to get something and that's what drove me. It was a great lesson that hard work pays off, and I got that message early.

## ADJUSTING THE APPROACH

Later, I found out that my hard work was not enough, that the quantity of time with the ball did not translate into quality. It wasn't enough to go out and just run or shoot, I had to work on my weaknesses and eliminate them. I wish I had watched my games on video and adjusted my workout better, but I did not want someone breaking my game down. I just wanted to go get pizza and forget about it, and that was a mistake.

## GOING NATIONAL

Even though I did not embrace the critics of my game, I did not run from the hard reviews, either. I remember our national coach, Anson Dorrance, saying, "Your defense sucks—it's really bad! Does anyone teach defense on the West Coast?" I responded well to this challenge and didn't melt down. I needed to improve my defense, so I improved my defense. With some women, even at the highest level, a critique could be taken to be personal, especially when it's delivered so directly, but it didn't bother me.

Anson still put me on the team when I was only sixteen, but I used to go to practice and think, "What the hell am I doing here?" Gradually, though, I showed the coaches and myself that I belonged.

### HOW IT ALL HELPS . . . AND STILL HELPS

When you go out in front of ninety thousand fans with the World Cup on the line you invariably learn about yourself and life. We won and lost the World Cup through my career, and each time I got something from the experience that helped me away from the game. For example, when I was president of the Women's Soccer Foundation, I took up the Title IX fight that helped give women the same opportunities as men when it came to college sports. I had to learn the debate backward and forward and the verbal sparring was exactly like a game, only I didn't break a sweat. I took on lawyers and was the proud and often only voice of dissent.

The road with the national team was so rocky for so long—seventeen years—that I learned to just scrap to survive. That's the mind-set I took to find out about the labor issues with Reebok. I just had to know under what conditions those soccer balls were being made, so I went to Pakistan and saw for myself that they were not using child labor. I approach broadcasting the same way today. I fight for the story and I always go all-out in the research. Why? Because that's the way I was trained to play soccer and that's how I have been trained to approach life.

### WHY WE ALL STAYED TOGETHER

One reason for our success can be summed up in one word: discipline. We all had focus, trained hard, and pushed each other. We would do anything to get to the top. We all had the same skills, the same tactical awareness, but we also had the mind-set and that's why we did so well, because we cared so much.

### HOW SHE DID IT—AND HOW THE TEAM MADE IT

We won and we stayed together because we played hard even when no one was watching. I was known to have an evil twin. I smile. I have fun. But I play hard. Very hard.

**MY WRAP**

*Julie was a great player with a great personality, someone who will do more with her life after soccer than she did before, because she is so driven, yet balanced. As a player, she wasn't flashy. She wasn't a scorer and she didn't care about publicity. But those who know her know how good she was. Her life is a virtual university on how to approach sports, and her soccer academies have adopted these principles.*

# MUHSIN MUHAMMAD

★ 2-TIME PRO-BOWL SELECTION—2000, 2005

★ 2004–NFL SEASON LEADER WITH 1,405 RECEIVING YARDS; LEAGUE-
LEADING 16 TOUCHDOWNS

★ NFL WIDE RECEIVER, CAROLINA PANTHERS, CHICAGO BEARS,
1996–PRESENT

You play the way you practice.

—POP WARNER,

*football coach*

I grew up in Lansing, Michigan, and there weren't any NFL players to come out of my city. In fact, I was the first one drafted and the first one to make it from Lansing. For me, going pro was anything but a done deal. I knew being a pro athlete was a possibility because Magic Johnson and John Smoltz both came from my town. But me? A pro? It wasn't on my mind. In fact, until I got to college, my favorite sport wasn't even football, it was soccer. And my second love was basketball.

When I got to Michigan State, I thought I would try to be a walk-on in basketball, but I messed up my knee and I thought it was over for me. Instead, it just pushed me toward football.

It was apt that football would be my ticket because the coach who made the greatest difference in my life was my Pop Warner football coach, David Miller. He exuded toughness. He had us doing up-downs till dark. That was a grueling drill where you hit the ground face-first and got up as fast as you could dozens of times, or until you were completely exhausted. I never even saw the ball because I didn't play a skilled position, which may be why I didn't love the game early on. Defensive end and offensive guard were my positions, and Coach Miller showed me how important they were. He always said the toughest players are the best players. He let me know I could be a really good football player, and because I didn't want to let him down, I worked my tail off.

## IT TAUGHT ME TO BE A TEAMMATE

I may be a pro now, but back then I wasn't even the best player on my team. In fact, until I reached high school I wasn't good enough to be the running back, quarterback, or a receiver. It didn't bother me that much, though, because I was trained as a team-first guy. I just went and made myself one of the hardest-hitting guys on the team. The coach made it fun and we won. Coach Miller is still around and he keeps in touch with me through my dad now. As soon as I signed with the Bears, coming from the Panthers, I sent him a signed jersey.

## FATHER KNOWS BEST

It was my dad who really foresaw my future as a wideout and he's the one who got me to play the position at Michigan State. It worked out, since I got drafted by Carolina and eventually went to a Super Bowl. Soon after that, I signed a free agent contract with Chicago that makes me among the best paid in the game.

Football helped give me confidence that I carry around with me today. I never say or think "I can't do this"—I guess you could say I have a warrior's mentality. It's just how I am. Even today, I still have the lineman's mentality because I don't seek the spotlight. It's not about me, it's about the team. I guess the best example of that is the 2005 49ers game where I caught one pass for eight yards and yet I got the game ball for the way I blocked. That was my role that Sunday, and it was okay with me as long as we won, and we *did* win.

## PERFORMING UNDER PRESSURE

I just love pressure. I love the big games because with the big game comes the big reward. The first time I felt real pressure was my senior year in high school. We were a smallish, Class A school and we were playing a powerhouse school for homecoming game. It was a packed house and I just managed to step up and score three touchdowns, one of which came on defense, and we won. Another game that stands out was when I was a senior at Michigan State. We were playing at home in front of one hundred thousand people and I caught a bomb over Charles Woodson—he wound up being a Heisman Trophy winner—and we beat them. I also managed to make some big catches for Carolina in the Super Bowl. I just taught myself to love those moments rather then fear 'em.

*He makes it sound easy, doesn't he? But what's he proudest of?*

That's easy. It has to be my charity work. I raise money and awareness for muscular dystrophy, a battered women's foundation, United Way, and the Boys and Girls Clubs of America. But the toughest and most rewarding job is, of course, raising my own four kids.

## WITHOUT THE GAME, DOES MUHSIN MAKE IT?

Without a career in the NFL, I think I would have been a successful businessman. I had good grades and graduated with a communications degree, so maybe I would have tried for TV or radio. I didn't think I had a shot at the pros until my junior year in college, but I already knew how to work and understood nothing would be easy. In fact, today, with my foundations, real estate, marketing company (Baylo Entertainment), radio show, and a TV segment on Comcast, I might be the busiest guy in the NFL. All of it takes a lot of my time, but when I retire I will have plenty of opportunities and companies to keep me going for years.

## MY WRAP

*Muhsin loves the game, but this is a man who did not need the game. He knows he has God-given talent, but he outworked many with more talent to become a feared, physical, elite wideout. Muhsin never stopped studying in school and never stopped thinking about business even after becoming a multimillion-dollar player.*

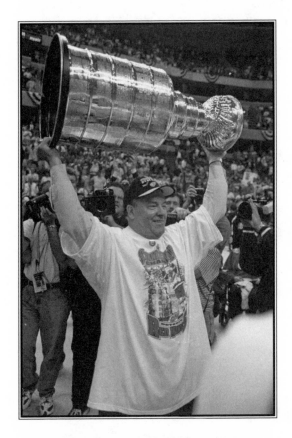

# SCOTTY BOWMAN

★ 9-TIME STANLEY CUP WINNER, 1973, 1976–79—CANADIENS;
   1992—PENGUINS; 1997–98, 2002—RED WINGS

★ NHL HALL OF FAME COACH, 1968–2002, ST. LOUIS BLUES,
   MONTREAL CANADIENS, BUFFALO SABRES, PITTSBURGH
   PENGUINS, DETROIT RED WINGS

If you can't beat 'em in the alley, you can beat 'em on the ice.

—CONN SMYTHE,

*original Toronto Maple Leafs owner*

I wanted to be a pro player but when I got a head injury at the age of seventeen, I was through as a player, though not with the sport. I wanted to get back out there, but the headaches would not stop. It was a down time for me, no question about that.

## MOMENT

I owe a lot to the Canadiens for what happened next in my life. They were the team I was looking to play for before my injury, and they still contacted me that summer. They offered to pay for my schooling if I worked for them, coaching some of their youth teams, so I did it. I was coaching about ten years before I finally got my first NHL post.

## TAKING A RISK

I was having some success at my day job in my early twenties, moving up the ladder at the Sherwin Williams Paint Company, when I got my first full-time offer to move into hockey. My mom and dad, coming from a blue-collar background, weren't for it because they had a hard time seeing sports as a job. But I didn't. I left after much deliberation and I have not regretted it a day since. Ironically, those days in paint helped with my coaching because the job demanded I memorize so many stats and paint types from so many different companies. My days in paint taught me to love numbers and statistics, which helped my coaching enormously.

## BOWMAN RULES: WAS THE YOUNG COACH BOWMAN ANYTHING LIKE THE LEGENDARY BOWMAN OF TODAY?

In some ways, sure, I was always into appearance. We had a dress code and I was a stickler for showing up on time. Both these qualities came from my own family. My dad demanded it and so I was used to it. I never missed a day of school and I don't remember my dad missing a day of work.

## TRY SOMETHING . . . ANYTHING

I always like to try new things. So what if it turned out goofy? Maybe some of these so-called goofy ideas will work. Besides, I always wanted to be unpredictable. It makes it more difficult for your opponent to figure out what you're going to do, and that can give you an edge. I also learned to embrace adversity, because for me, success was right behind it. I've always believed that you are only as good as your last game. The good coaches and players who win the Stanley Cup don't carry their rings around with them. They know better. You win. You're happy. And then you become a target.

## PROUD

The thing I'm proudest of is that I've lasted five decades. That's a long career and it turned out okay. Funny thing is, when I began, people said I was too young. In the end, they were saying I was too old!

I get a thrill when I learn that my former players went on to do well off the ice. It's great because it means I may have taught them something positive, something they can use after their hockey careers have ended.

## HOW SCOTTY PLAYS THE GAME

Do your best and the worst won't happen.

## MY WRAP

*For a life dedicated to hockey, Scotty had so much else going on. His parents instilled in him such great core values that he would have been good at anything he tried. The game never flew by him because he kept ahead of the curve with his dedication to innovation. That's why his teams won so many games and so many Stanley Cups. Personally, to see him put on the skates and whiz around the rink with the Cup in Detroit showed why he claims he never worked a day in his life. It's because he loved what he did for a living. Notice, too, that he took a chance when he left a secure job to focus on his passion. Clearly it was a risk worth taking, but how many others, I wonder, wouldn't have taken that leap?*

# ROBERT NARDELLI

★ SELECTED BY PRESIDENT GEORGE W. BUSH TO SERVE ON THE
   PRESIDENT'S COUNCIL ON SERVICE AND CIVIC PARTICIPATION

★ FORMER BOARD OF DIRECTORS, THE COCA-COLA COMPANY

★ FORMER CEO, HOME DEPOT

★ FORMER SENIOR VP, GENERAL ELECTRIC

The road to the boardroom leads through the locker room.

—DAVID RIESMAN,

*sociologist*

Sports were always a big part of my family life. It was just what we did when I grew up. My brother played football at Purdue and went to the Rose Bowl, so it was natural that I would play football, too.

My first uniform included a gold leather helmet without a face mask. I remember freshman year at school, at orientation, one of my teammate's hands shot up and he said, "Where do I get my mouth guard?"

The coach said, "Boy, you still got all your own teeth?"

That set the tone for my college career, and it let me know how the coaches wanted the game played.

Football taught me a lot about myself, and at the same time it taught me about teamwork and dependency on others and their dependency on you. Every play was designed to go for a touchdown. It was only when there was a breakdown in the assignment, or lack of communication, or an opponent simply outmaneuvered us, that the plan failed and doesn't go all the way. That concept is a basis for life in and outside the boardroom.

I was 5'10", one hundred ninety-five pounds, which made me the smallest lineman on the team for all four years. Nevertheless, I didn't miss a start in all that time. Being a lineman, I learned a lot about accountability. I felt like I was being watched every moment of every game and, in fact, I really was. I was rated each play every play and I would get a score, which was posted every Monday morning in the locker room. This is what taught me accountability early on. It was made clear to me that if you deliver you start, which really helped me in my early career in business.

**LEARNING TO VISUALIZE**

Coach had everyone on the team go through every play in our minds, telling us to ask ourselves, *How will you handle everything your opponent might do?* This allowed us to live the game before the first snap, and it really worked for me my entire college career. At my size, I had to outthink my opponent every game, every play. I use this technique today in business. Before giving a presentation at a big meeting I ask myself, *What could go wrong, what questions might I get?* Doing this gives me confidence that I'll be ready for anything. It makes me sharper and lets me think proactively, allowing me to be the predator rather than the prey.

## OVERMATCHED? SO?

I remember there was one guy who I locked up with who was like a cement wall, a real freight train, one of the best players on the Hilltoppers of Western Kentucky. When we walked into the locker room coach had "The Impossible Dream" playing on the sound system. We got the message: we were going have a big challenge that week. What I had to do was use my speed against my opponent's raw size and power. I'm right-handed, but I would get down in my stance with my left arm down, which kept my power arm available to deflect a linebacker coming at my head, hoping to run me over. How did it turn out? We lost and I took a physical beating—I guess we all did—but I was proud that we battled through and never took a play off. I tried everything I knew and it was not enough, but I have no regrets. I knew then that I could give everything you've got and still lose, but you can still take pride in the effort.

## PERSONAL PRIDE

I relish the fact that the team voted me captain to this day. Even now, looking back, I'm moved by it. I'm also proud that I found a way to stay in the starting lineup as the team got bigger around me. I'm especially proud that our Western Illinois team was ranked nationally and that many of our players hold individual team records that still stand today. Most importantly, the team got better after we left, and I think our group had a lot to do with that. We set the standard and stayed in contact with the program.

*That was then. Tell me about now.*

The lessons I learned playing football were portable. As much as I learned and enjoyed playing football, I had to move on. This isn't Hollywood. Each day I had to prove myself at Home Depot and before that, General Electric. I have a core purpose and that's to improve everything I touch.

**PRINCIPLE TESTED**

There's no question that I wanted to be the next CEO of General Electric after Jack Welch left and I was doing everything I could to make it happen. I worked in so many divisions within the company that I put myself on the short list to be the new CEO. I refused to think about a contingency plan or plan an exit strategy should I not get it, because it would have been a distraction. Well, Jeff Immelt got the nod and Jack's explanation was simple: he went with his gut. I allowed myself a short time to be disappointed and then it was on to my next challenge. Thankfully, it was a great company, Home Depot. Football taught me that you don't win every game, or in this case every job, but what you can do is take some lessons from your experiences and then move on.

*Now that Bob has left Home Depot, I know those principles and his resilience will once again be put to the tests and I for one would never bet against him running a major corporation again. He is also in a position, unlike most of us, to say "Game Over" and retire to the Pro Bowl of life.*

**MY WRAP**

*Bob's early sports experiences prove how sometimes you have to take a series of blows, but that doesn't mean you still can't come up a winner. I just hope Bob doesn't find out that I also have Jeff Immelt in this book and that I had Jack Welch in my first!*

*Medavoy, second row, third from right*

# MIKE MEDAVOY

★ COFOUNDER, PHOENIX PICTURES, 1995: *THE THIN RED LINE, PEOPLE VS. LARRY FLINT*

★ CHAIRMAN, TRISTAR PICTURES, 1990–1994

★ FOUNDER, ORION PICTURES, 1978: *PLATOON, TERMINATOR, AMADEUS, SLEEPLESS IN SEATTLE* (EXECUTIVE PRODUCER)

★ UNITED ARTISTS, SENIOR VICE PRESIDENT OF PRODUCTION, 1974–78

There is no substitute for hard work and effort beyond the call of mere duty. That is what strengthens the soul and ennobles one's character.

—WALTER CAMP,
*Yale football coach*

I started playing soccer when I was seven, in South America, and it was all-consuming. Weekends I was on a team, and during the week I played in the neighborhood with twenty of my buddies. No leagues, no drills—we just played. Part of the reason I liked it so much is that I stood out. Even when I came to the United States I played at UCLA, where our team had a great deal of success.

Sports helped me blend in, and my first year here I was quick to join the tennis and swim teams, along with the soccer team. It was a time, in the 1960s, when very few of my teammates were Americans. All of us were new immigrants, many of them here only temporarily. I was here to stay. I kept playing through my fifties and I can still handle the ball, even today.

## EXCEEDING EXPECTATIONS

I scored a lot of goals as a kid at the ages of ten, eleven, twelve, and what made it special was that no one expected me to be a scorer. One time a kid just walked up to me and said, "I'm a better player than you are." I was taken by surprise and I didn't say anything back. But something clicked inside me and I scored seven goals against his team. Afterward he came up to me and said, "I was wrong."

## UCLA SUCCESS

In the four years I was at UCLA, we only lost one game. The match that sticks out was the one game I played halfback. I got injured and it was the first injury I'd had that kept me out of action. It was at that time I realized how much I loved the sport. It was significant to me because I never quite recovered from the ankle injury. I came back too soon and I wasn't the same player.

As a tennis player, I was good enough to make the team, not great. But with soccer, I was one of the better guys on my team. The reason was simple: I stood out because I learned the game and acquired the skill in Chile, and the level of play in the United States was not as good.

## NO ILLUSIONS

I had a sense I was a good player, but never good enough to go pro. I got my lesson on my limitations when I one day took the field against a guy who eventually would become captain of the Chilean national team. He just destroyed us, and I couldn't touch him. That's when I knew I was good but not good enough. Besides, I always wanted to do different things, not just play soccer. I always looked for roundness in life. I never was about doing anything perfectly well. I was about doing a lot of things as well as I could. I would work on my game, but I did it because I enjoyed it, not because I had too.

## TRAIT SECRETS

I had a good sports instinct, especially in soccer. That might be my greatest asset. I could anticipate where a ball would be played, where a player would be, and I like to think I brought that to the movie business. That talent helped me work with people in the entertainment business on many levels. What I liked about playing the game is I could see the field and had the skills to make other players better. That's exactly what I do now. It's not about me scoring, it's about the team winning. The movies I make are never about me, it's about creating the best movie possible.

I think people resent those who try to take the credit or inflate their importance. If the art director is lousy, it will affect the movie. If one of the actors is lousy, it will affect the movie. If you have the wrong player, or a player in the wrong position, you will lose the game. Remember, Brazil has won more World Cups than anyone else, but the years they lost it was because they didn't have a strong defense and they tried to make up for it with a superstrong offense. It didn't work. And it never does.

## HOW I MAKE MOVIES

I approach movies as an endeavor that leaves a mark. I have been accused of making movies that are too good. And maybe it's true. I tried to make too many classy movies and not enough commercial movies. I approach everything from the team perspective, and I've always been about doing things right and well. It helped me to make sports movies like *Eight Men Out, Bull Durham, Hoosiers, Rudy*, and *Caddyshack*. As an athlete myself,

I could watch and see what rang true and what didn't. Being an athlete helped me when I read the script. But more than just knowing about sports, I know about human nature, which is about how we behave in pressure situations, how we handle winning and losing.

### ABOUT WINNING

Whether it's a movie or anything else I do, it's all about trying my hardest and sticking to it. If I've done all I can do for any project, then that's enough for me. My parents wanted to make sure I had the survival instinct in me. Some people might like me, others might be jealous, some might think I'm not a good guy. But I'm in my own skin, and that's pretty okay.

### WHAT I WISH

I never had that extra step as a player, and the older I got, the more endurance I lacked. I found out why recently: I was living with a heart valve problem. It affected my endurance, and I wonder how much it affected me in my younger playing days. I also never had a coach who did more than put the players into positions. I never had that guy who drilled us in fundamentals. My son will have that and I know he's the type of kid, like me, who will really absorb it.

### THE GAME

In 1980, I remember playing at the Meadowlands (Giants Stadium) before a Cosmos game. I was forty years old, playing in a New Jersey all-star game. I stuck around and watched the pros afterward and I saw how much better they were through the eyes of a forty-year-old.

### FINAL THOUGHTS

You approach every game and everything else in life by saying *I want to win*. You ask yourself what will that take and your answer should be, "I have to play my best and my teammates have to do their part. When everyone does his or her best we either win because we play well or lose because we are not good enough." That's it. You go hard, max out the effort, and deal with the results.

**MY WRAP**

*Mike, like so many others in this book, lowballs his skill, accomplishments, and success. You get the sense in talking with him that it's not about seeing his name in the credits or on the stat sheet. He has supreme confidence, but he's not cocky and he's only interested in results, not a pat on the back. Isn't that the definition of a team player, a great captain, or a great coach? I'll answer that one for you: Yes!*

# JOE MONTANA

- ★ **NFL HALL OF FAME, 2000**
- ★ **NFL QUARTERBACK, 1979–1994, SAN FRANCISCO 49ERS, KANSAS CITY CHIEFS**
- ★ **2-TIME NFL MVP, 1989, 1990**
- ★ **3-TIME SUPER BOWL MVP, 1981, 1984, 1989**
- ★ **4-TIME SUPER BOWL CHAMPION**
- ★ **NCAA NATIONAL CHAMPION WITH NOTRE DAME**

I've been playing this game for eighteen years, and I haven't yet figured a way to get into the end zone when you're on your rear end.

—FRAN TARKENTON,
*former NFL quarterback*

## HOW THE BEST ALMOST BAILED FOR THE BOY SCOUTS

I owe so much to my dad because he never let me quit anything. I learned the hard way because I was always looking to take the easy way out and he wouldn't let me. The closest I ever got to quitting football was at the age of eleven. I hated doing the sprints and the long practices and I saw what a great time my friends were having at Boy Scouts. So, one day, I thought I'd just quit football. I went home, found my dad, and told him about my exit plan and believe it or not at first he said okay. But then he thought better of it and said, "Make it through the season and then make your decision. You made a commitment and you've got to keep it." I ended up having fun that year and decided to stay with it, and it worked out for me.

*I'd say so, as would Notre Dame, the San Francisco 49ers, and the Kansas City Chiefs.*

Looking back, I don't know why I wanted to quit, because my team had a lot of success: I guess I may have just been tired of the training because I know I always loved the game. But my dad sent me a great message that I stick with to this day: don't ever quit something you start.

## WHY I HAD THE CONFIDENCE

If you have success you gain confidence, and in sports I was always near the top in my age group. Even if someone was better than me, I would never admit it. They would have to prove to me that their success was not a fluke. I would keep playing that person in whatever it was, mostly basketball at that time, until I beat them. By not giving in, I forced myself to raise my game to prove myself correct.

I wanted to be the person who would do whatever it took to win, and I was always convinced that winning was more then a possibility. I also learned to keep all my efforts in the game right here, right now. I learned never to look to the next game or the next season, only the next play. In sports, worrying about what's going to happen next is usually a total waste of time and just gets in the way. This is what helped me keep composed on the field at all times. Sure I got nervous and lost confidence, but

I was also good at hiding my emotions. I would never let my emotional state effect my ability to function. I didn't fear being in the game, playing quarterback, or making split-second decisions. I craved it.

*Which is probably why he was called Joe Cool.*

### THE FIRST SIGNS OF COOL—PERFORMING UNDER PRESSURE

I was up for starting quarterback in high school and had a tough fight in front of me. But I knew I was better and I won the position. Well, this guy was ticked, and what made it tough for me is they switched him to defensive back and he just hammered me every day in practice. I didn't complain. I didn't like it, but I rolled with it because that's what I learned to do from my dad. I always wanted the ball. I always thought, *"I can win,"* and that personifies what my dad drilled into me.

### THANKS, DAD

This intensity and the will to win came from my dad. No doubt about it. My dad and I used to play basketball, and since I was quicker I'd always blow by him and he would always trip me up. A clear foul! I'd say, "You can't do that!"

And he'd say, "I just did it, what are you going to do about it?"

Although I was seething, I'd keep playing, and then he would step on my foot when I went for a rebound. I'd complain and he'd laugh and say, "What are you going to do about it?" It showed me that during a game, things are going to happen that might be wrong, but they're out of my control. This taught me to deal with any obstacle and fight through it.

### WAKEUP CALL

My dad didn't stress academics, and I paid the price for that in college. I was in shock early on. Thankfully, I was taken in by one of the coaches and he threatened me to make sure I went to class. Some of my teammates would take me to their homes on weekends, and it all helped me get the input I needed to get by at Notre Dame.

### AFTER THE GAME

I was involved in sports from the age of nine. After the game was over for me I did have trouble making the transition into retirement. I never considered myself a very good spectator, but I never realized how much I would enjoy watching my boys grow up and play sports. I am much more nervous watching them play than I ever was as a player. I missed the girls' early years because I was always playing, and I was determined to handle the boys differently. I learn so much by watching them today and I hope I'm helping them the same way my father helped me.

## HOW HE PLAYED THE GAME

I would describe the way I played the game, any game, as scrappy and prepared. I prided myself in doing what I needed to do to get it done, and I was successful at it. I was never the biggest or the strongest, nor did I have the best arm, but I found ways to get it done. And even though I got a lot of the attention, I never put myself ahead of the team. Maybe that's why so many of my former teammates are my friends today.

## MY WRAP

*You don't have to be the biggest, the strongest, even the best, to get it done. You just have to be prepared, like Joe, to give everything you have and, maybe more importantly, be prepared to deal with the unexpected.*

# BRIAN KILMEADE

★ COHOST, *BRIAN AND THE JUDGE,* FOX NEWS RADIO'S NATIONAL NETWORK, 2006–PRESENT

★ COHOST, *FOX AND FRIENDS* (FOX NEWS CHANNEL), 1997–PRESENT

★ COACH, MASSAPEQUA CELTICS BOYS SOCCER, 2006–PRESENT

Some people think football [soccer] is a matter of life and death—I can assure them it is much more important than that.

—BILL SHANKLY,
*Scottish soccer manager*

My organized sports career consisted of one sport for a lot of years—specifically, soccer, which I played from the age of five to twenty-two, and still coach today. I was the kind of player who took it very seriously, with mixed results. I played throughout college on excellent teams, but few would walk by any game I was playing in and say, "If we could only stop number nineteen, we'd win easily." I was more the defensive midfielder/outside back, with below-average speed and above-average endurance. Now that I've done everything I can to convince you to skip over my saga, I'll tell you two stories you probably can relate to, stories that still affect me today.

One happened in seventh grade, while playing for an undefeated CYO team in the Catholic league finals against St. Patrick's. They were also undefeated and the first team to really play us dead even all season. With time running out and me playing outside back, a St. Pat's player with lightning speed broke free, beat the opposite side back, then the center back, leaving me having the choice of leaving my man and picking up the ball or just letting him go in on the goalkeeper. What do you think I did? I went for the ball, because that could hurt us the most. He was driving up the middle, saw me coming, spotted my man open, and fed him the ball. He shoots. He scores.

From the opposite sideline, in front of fifteen hundred spectators, I hear my coach yell in his Scottish twang, "Brian, that was your man and you just cost us the championship."

Nice! Here I am thinking I'm being unselfish and I get that rip. What a great feeling to have in front of my family, friends, teammates, strangers, opponents, janitors, and future book buyers. Well, my dad, who was coming straight to the game after a night of working in his bar, was in no mood to see his son verbally undressed and almost tore my coach's head off. The coach later apologized. I accepted, but never forgot.

## WHAT I LEARNED

Up to that point, I was under the impression that adults were always right, especially coaches. But that experience taught me that adults aren't infallible. I knew the best thing for the team was for me to leave my man and pick up the guy who could hurt us the most. I was stunned to be called out like that, not to mention the sting of suffering our first loss in seventeen games. Looking back, I also began to hate the big game because it meant the possibility of a big mistake.

The embarrassment lingered for years. Now, as a coach myself, I always look to smooth the fall of any kid who's singled out for criticism, warranted or not. I never point a finger and will get on any player who does. This has carried over to my career in broadcasting. When it comes to a big sports or news story, I thirst for the ball. If I'm prepared, it usually works out. If I struggle, I deal with it, because it's all about the effort you put in, not just about the victory.

## LAST STORY—I PROMISE

Before Day One of my freshman year in college, I found myself on the roster and likely to be the first off the bench on a real strong, veteran Division II soccer team at C.W. Post College. My coach was so confident of the strength of our team that he arranged for our first three games to be against three of the best teams in the country, including the University of Southern Florida, a Division I power. We opened up slow, falling behind, because no one could contain a guy named Roy Wegerle. He was a forward who went on to play the next twelve years on the national team and whose brother was a star player on the Cosmos. My coach, George Vargas, stuck me in at outside back, to mark him. I thought he was crazy because here I was, a player who had trouble cracking the lineup in high school, and I was tasked with running down the most talented player I'd ever seen play, let alone tried to cover.

I held him scoreless for the rest of that game. I was fifteen hundred miles from my family, friends, and hometown, and with all expectations lifted, I did incredibly well. He didn't score and I don't think I made a bad play in fifty minutes. The next game, I was the subject of the coach's pregame speech and I got the start against the reigning Division II national champs, the University of Tampa.

## MY WRAP

*What I learned from what for some might seem to be a relatively insignificant sports moment was that I created my own pressure and expectations. How I processed that pressure would ultimately determine my performance. I had never felt so free on the field in my life because I knew no one was judging me and I wasn't judging myself. After making the roster, I felt like I was playing with house money. No one in my family, nor any of my friends, ever put pressure on me. I put pressure on myself, because my self-esteem was tied into what I did on the field. If I played bad, I was bad. If I played good, at least for a few days, I was good.*

*I know it sounds crazy, but it's how I felt and I think it's how a lot of people feel. I came to learn that pressure comes from within, and how we handle that pressure defines how we do later in life, as well as the degree to which we enjoy the process.*

*Now I can get inside my head before going on TV or radio, and nine times out of ten, I perform well under pressure. I don't bring my self-esteem to the set, but I do bring my soccer work ethic. I don't know if I would have these life skills if I hadn't at least laced up the shoes and gotten into the game. I look back at much of my approach to sports and do the opposite, which worked for many in this book. It certainly works for me.*

# AFTERWORD

To put it simply, if you are a parent, player, or coach and I have not influenced you on how to approach "the game" after reading this book, I've failed. Personally, I don't see how you can sit down with these stories and not find plenty of clues to leading a successful life. I know it's a cliché, but nevertheless, in the end, like me, I hope you will conclude that it's not whether you win or lose, but how you play the game that counts.

In case you hadn't noticed, almost all the athletes had one person who profoundly impacted not only their game but their life. Whether it was Sean Elliot taking to heart his sister's boyfriend's review of his skills, or Rush Limbaugh getting a tongue-lashing for pacing himself when he should have been gutting it out on the gridiron, or football player John Lynch being told by his basketball coach that he could be special because he knew how to compete, the ultimate impact was the same. They started to believe that their approach to their game was wrong and had to be adjusted. Later, this lesson carried through to their lives. Coaches should ask themselves this question: Am I going to let my superstar know how good he could be if he worked hard (like Elliot) or let my player dog it day in and day out because he's not a star (like Rush) or not fuel the fire of my blue-collar star?

All the people in this book had another thing in common: they dealt with setbacks not by ignoring them, but by tackling them head-on. For Steadman Graham, his intolerant on-court attitude embarrassed him and his team, and as a result of that effort to change, he became one of the most successful businessmen around. Joey Cheek despised losing so much that he started to hate skating. But rather than quit, he adjusted his attitude, attained Olympic success, and, now that his sports days are behind him, has set a new standard for benevolence. Troubled waters and turbulent times were an absolute necessity with all these athletes and so the lesson to parents, as hard as it might be, is that maybe you should let your kid deal with his benching, that unkind cut from the team, or that unfair coach without stepping in. Instead, let them handle it when they're young, rather then experience it for the first time in college, business, or marriage.

Another theme common to all these athletes, and one that you've certainly heard before, was sacrifice. Mia Hamm was already on the national team when she learned that she really didn't know what hard work meant. She went on to sacrifice her summers and all her free time for almost ten years to achieve her lofty status. Rebecca Lobo had a similar experience at UConn. As kids, Mark Brunell and Joe Montana both talked about the agony of giving up their vacations for sports. They were pushed to not quit, and neither of them have any regrets. Elisabeth Hasselback sacrificed summer months to training to be a pinch runner and eventually became captain of her softball team as a walk-on. Are you going to tell me that didn't lay the ground work for shining on *Survivor* or winning the host job in the TV jungle on *The View?*

Regardless of who you are, what era you played in, what sport you chose, or how much success you achieved, playing the game is all about getting you ready for life. Winning or losing has little to do with who you will become. Instead, it's how you prepared for the game that determines whether you'll be a winner or loser in life. Having said all this, the only question that remains is: How do you play the game?

# PHOTO CREDITS

James/Dallas Stars; 167, David Hume Kennerly; 170, University of Georgia; 174, courtesy of Beth Ostrosky; 177, Theodore Roosevelt Collection, Harvard College Library; 181, Virginia Tech Athletics; 185, The University of Arizona Athletic Media Relations; 193, courtesy of Gale Sayers; 198, Doug Pensinger/Getty Images; 202, photos by Doug Pensinger/Getty Images; 206, courtesy of Martin Jacobson; 209, HPG-Mickey Holden; 212, CBS; 216, TK; 220, photo by Jed Jacobsohn/Getty Images; 224, Fine Family Archives; 227, Joe Traver/Time Life Pictures/Getty Images; 232, University of Illinois Photo Department; 236, Lincoln Presidential Library and Museum; 239, courtesy of Vincent O'Connor; 243, University of Virginia; 247, courtesy of Ben Nighthorse Campbell; 254, courtesy of Van Colley; 257, courtesy of Harvey Mackay; 261, photo by Brian Bahr/Getty Images; 267, Art Seitz; 271, photo by Focus on Sport/Getty Images; 274, courtesy of Hank Williams Jr. Enterprises; 277, courtesy of Bowdoin College Archives, Brunswick, Maine; 280, from the Frank Kracher Collection, National Soccer Hall of Fame, Oneonta, NY; 283, courtesy of Tony Siragusa; 286, courtesy of Mary Lou Retton; 290, courtesy of David Brunell; 294, courtesy of Gary Player; 299, Bill Wood; 310, UConn Athletic Communications; 314, courtesy of George E. Foreman; 318, USC Sports Information; 321, photo by Doug Pensinger/Getty Images; 325, photo by Rusty Jarrett/ Getty Images for NASCAR; 329, courtesy of George Shultz; 332, photo by Jamie Squire/Getty Images; 336, photo by Ezra Shaw/Getty Images; 340, Rick Stewart/Allsport/Getty Images; 343, courtesy of Robert Nardelli; 347, courtesy of Mike Medavoy; 352, Joe Raymond; 356, courtesy of the author.

# ACKNOWLEDGMENTS

Special thanks to Cal Morgan for executing the game plan, and to Bob Barnett for making sure it all worked. Thanks also go to my editor, Doug Grad, for his incredible attitude, sports knowledge, and coaching, along with Sarah Haley for tirelessly helping hunt down those unique photos. Thanks to Judith Regan for making this second book a reality. I'd also like to thank Charles Salzberg, a terrific writer, teacher, and sports fan.

This project and so much more is all because Roger Ailes kept me around the world's best news channel, Fox News, for ten years. Who can ignore Fox News Channel vice president Bill Shine for his unending faith and VP Kevin Magee for steering me through TV and now radio as well? Did I say radio? Patiently working around my book writing schedule is the medium's A team—Mike Elder, Jeff Hillery, Sean McGrane, Griff Jenkins, and Joey Salvia. Of course, working with Judge Andrew Napalitano as my radio partner is like working alongside a living, breathing institution of higher learning each day. The judge is also my mom's favorite on-air talent. On *Friends,* who could have better co-hosts then Gretchen Carlson and Steve Doocy? In the case of Doocy, I can only imagine how intolerable it must be to put up with me for three hours a day for ten years, but for me it's been a thrill. Of course, E.D. Hill knows all about my taxing style from the eight years we were together. Who cannot spotlight my *Fox and Friends* first-team all-stars Kiran Chetry, Lauren Green, Alisyn Camerota, and this guy named Tiki Barber. Of course, helping me every day and supporting this book in every way is the *Friends* production staff led by executive producers David Clark and Jerry Burke, senior producers Maria Donovan, Gresh Striegel and Christine Thoma, and our team, Laurie Weiner, Jose Lesh, Gavin Hadden, Paulina Krycinski, Tiffany Fazio, Marcelo Garante, Monique Dinor, and Jack Savage. For helping me make the contacts, my esteemed thank-yous go to Ron Messer, Jennifer Williams, Joe Krause, and the legendary Craig Thomas.

If I had a Founder Award to issue, it would go to Jack Abernathy, a Fox executive who thought my first book would soar and was equally supportive of this one. Who knows—maybe he'll turn it into a TV series. He truly has that much power

You don't book this many interviews and gather pictures from ninety people without a lot of help from behind-the-scenes stars like Nova Langtree for grabbing me three of her clients; Joel Segal for doing the same; Dick Lynch, a great friend and the American military's best friend, Kristi McCormick for her unending quest to get me the best and brightest while always cheering me on; Megan Henderson for always looking out for me and the book; Jean Osta-Niemi; and Kim Shreckengost for delivering me the business titans Arthur Blank and Robert Nardelli. Thanks go to Leigh Steinberg, who opened up his Rolodex for me. I am deeply honored. Bob Ferraro and Fred Engh have been great resources, always making themselves available along with providing moral support and allowing me to speak to their incredible youth sports organizations.

Other VIPs: Blake Bandy, Gary Sheffer, Craig Miller, Andrea Ross, John Maroon, Cori Britt, Meredith O'Sullivan, Jessica Attebury Quinn, Jeff Quinn, Charles Sherman, Jennifer Mayfield, Ty Norris, Van Colley, Greg Baily, and John Lynch Sr.—a proud dad and a great boss. Still more MVPs: Kirt Webster, Claire Everett, Tammi Olsen Starr, and Patrick Goldberg.

This book, unlike my first, includes history's greats and many secondary sources. The following deserve praise for their time and insight: Doug Wead, Jamie Totten, Joe Meyer, Jonathan Aitken, Eric Lamond, Wallace Dailey, John Staudt, Jim Foote, Tweed Roosevelt, Dennis J. Hutchinson, Jerzy Kluger, Gretchen Wayne, and Tim Kelly.

Of course, I would not even have had a book if all these wonderful successful people had not given me quality time in the midst of their incredibly busy lives.

The following people shaped me and helped define this book by what they gave me while I was on the field as an athlete. First off, the incredible soccer coaches Frank Fergus and Carl Knoblach, who I later found out would trade for me every year and always found an eager partner. I will never forget the late Walter Uschok for putting me on his team and most of all for asking my dad to be his assistant, launching my dad's own youth travel coaching career. Thanks to George Vargas for making me a two-year starter, including giving me a shot to match up directly against my brother as a freshman. No one meant more to me on the sideline then my brother and dad, or my mom who never stopped supporting me. They let me know that effort and drive were everything. Most of all, they let me play. And to all the coaches who chose not to start me, play me, or pick me, I thank you even more because it let me know that life's not fair, that I had to suck it up and not quit, and learn that things would eventually work out. Although it took twenty years, things eventually did work out!

Of course, sports supplemented all I learned at home, with my parents not only showing me how to approach life but also how to treat people. Thanks, Jim, for being that guy any younger brother would love to have to look up to. Thanks, Steve, for being a guy whom you would want to have your back in a bar fight, to fix your car in a snowstorm, and to outperform your stud employee in any workplace in the country.

Last and best, thanks to my mom and dad, who I live to not let down.

I like to think that the players I coach today, especially the ten-year-olds on the Massapequa Celtics, realize that I know the sub is just as important as the star, and that it's not whether you win, but it's how you play the game that matters.